ENVIRONMENTAL PROTECTION IN TRANSITION

For our wives, Susan Erickson and Izabela Kowalewska-Cole

Environmental Protection in Transition

Economic, Legal and Socio-Political Perspectives on Poland

Edited by
JOHN CLARK
Hudson Institute

DANIEL H. COLE
Indiana University School of Law Indianapolis

Routledge
Taylor & Francis Group
LONDON AND NEW YORK

First published 1998 by Ashgate Publishing

Published 2017 by Routledge
2 Park Square, Milton Park, Abingdon, Oxfordshire OX14 4RN
711 Third Avenue, New York, NY 10017, USA

First issued in paperback 2017

Routledge is an imprint of the Taylor & Francis Group, an informa business

British Library Cataloguing in Publication Data
Environmental protection in transition : economic, legal
and socio-political perspectives on Poland
1. Environmental protection - Poland 2. Environmental policy
- Poland
I. Clark, John II. Cole, Daniel H.
363.7'09438

Library of Congress Catalog Card Number:
Environmental protection in transition : economic, legal and socio
-political perspectives on Poland / edited by John Clark, Daniel H.
Cole.
p. cm.
Includes bibliographical references and index.
ISBN 1-84014-422-X
1. Environmental policy–Poland–Congresses. 2. Sustainable
development–Poland–Congresses. 3. Poland–Economic policy– 1990-
–Congresses. I. Clark, John, 1958- . II. Cole, Daniel H.,
1958- .
HC340.3.Z9E525 1998
363.7–dc21 98-6528
CIP

ISBN 13: 978-1-138-27985-8 (pbk)
ISBN 13: 978-1-84014-422-2 (hbk)

Contents

List of Tables *vii*
List of Contributors *ix*
Preface *xii*

1 Poland's Environmental Transformation: An Introduction
Daniel H. Cole and John Clark 1

2 Sustainable Institutional Design in Poland:
Putting Environmental Protection on a Self-Financing Basis
Zbigniew Bochniarz and Richard S. Bolan 19

3 Economic Instruments in Polish Environmental Policy
Jerzy Śleszyński 30

4 Privatizing the Polish Economy:
Benefits and Dangers to the Natural Environment
Bogusław Fiedor 68

5 Environmental Law in Poland, 1989-1996:
An Assessment of Past Reforms and Future Prospects
Jerzy Jendrośka 81

6 Environmental Protection and Mass Privatization:
A Management Perspective
Susan Cummings 116

7 Polish Greens and Politics:
A Social Movement in a Time of Transformation
Piotr Gliński 129

Index *154*

List of Tables

Table 1.1	Percentage Changes in GDP, Industrial Production, and Air Pollution Emissions in Poland, 1990-5	4
Table 3.1	Major Distribution Schemes for Collected Environmental Fees and Fines	33
Table 3.2	Revenues Provided by Environmental Fees and Fines	35
Table 3.3	Revenue from Environmental Fees and Fines in 1994	35
Table 3.4	Emission Fees for Atmospheric Pollutants	39
Table 3.5	Fees for Waste Water	40
Table 3.6	Fees for Surface Water Withdrawal	41
Table 3.7	Fees for Ground Water Withdrawal	41
Table 3.8	Fees for Waste Disposal	42
Table 3.9	Tax Differentiation on Gasoline	42
Table 3.10	Deposit-Refund on Beverage Containers	43
Table 3.11	Revenue Sources for Environmental Investment	46
Table 3.12	Environmental Expenditures by Sector	46
Table 3.13	Total Environmental Investment	47
Table 3.14	Environmental Investment Indicators	48
Table 3.15	Environmental Expenditures in 1993 by Sectors	48
Table 3.A1	Emission Fees for Sulphur Dioxide	62

Table 3.A2	Emission Fees for Nitrogen Oxides	62
Table 3.A3	Pollution Fees for Waste Water	63
Table 3.A4	Pollution Fees for Saline Water (Coal Mining)	64
Table 3.A5	Fees for Surface and Ground Water Withdrawal	65
Table 3.A6	Fees for Waste Disposal	66
Table 4.1	Privatization and the Principles of Sustainable Development	77

List of Contributors

Zbigniew Bochniarz is Visiting Professor and Director, Central and East European Environmental Research and Training Project at the Hubert H. Humphrey Institute of Public Affairs at the University of Minnesota. In addition, he is a member of the Advisory Council of the Institute for Sustainable Development in Warsaw. He has written or edited several books, articles and studies on environmental policy and sustainable development in Poland, Hungary and other countries of Central and Eastern Europe.

Richard S. Bolan is Professor and Director of Planning at the Hubert H. Humphrey Institute of Public Policy at the University of Minnesota. He has written or edited several books, articles and studies on environmental policy and sustainable development in Poland, Hungary and other countries of Central and Eastern Europe.

John Clark is Director of the Hudson Institute's Center for Central European and Eurasian Studies and Adjunct Professor of Political Science at Butler University. Co-author (with Aaron Wildavsky) of *The Moral Collapse of Communism: Poland as a Cautionary Tale*, he also served as an advisor to the democratic opposition in Poland and Hungary during the late 1980s.

Daniel H. Cole is Professor at the Indiana University School of Law at Indianapolis, where his subject areas include Property Law, Environmental Law, Natural Resources Law and International Environmental Law. He is a recognized American expert on Polish environmental law, having written several articles and a book on the subject. Professor Cole holds a J.D. degree from the Northwestern School of Law of Lewis and Clark College and a J.S.D. degree from the Stanford Law School.

Susan Cummings spent several years in Poland as a legal and economic advisor on environmental issues in the privatization process. In 1992 she assisted Poland's Privatization Ministry and the Ministry of Environmental Protection, Natural Resources, and Forestry in creating the Interministerial Environmental Unit (IEU), which resolves environmental issues in privatization transactions. From 1992 until 1994, she served, first, as Advisor due diligence in preparation for privatization and joint venture transactions. From 1995 through 1996, she served as a Fund Manager for Creditanstalt SCG Fund Management, which manages a portfolio of 34 companies participating in Poland's mass privatization program. Ms. Cummings also has published several book chapters, law review articles, and editorials on environmental issues in Central and Eastern Europe. Ms. Cummings holds a J.D. degree from George Washington University National Law Center.

Bogusław Fiedor is Professor and Head of the Economics Department at the Oskar Lange Academy of Economics in Wrocław, Poland. He has served as a consultant to the World Bank Environment Mission to the Russian Federation, and for the Bank's Industry and Energy Department, Environment Division. Professor Fiedor has worked with the Harvard Institute of International Development and the World Resources Institute, as well as with the energy divisions of the European Bank for Reconstruction and Development and the United Nations Conference on Trade and Development. He serves as an expert advisor to Poland's Parliament on environmental economics, a subject on which he has published more than one hundred books and articles. He has written with particular emphasis on the transformation from socialist to free market economics, ecological constraints on economic growth, Poland's economy-wide environmental policies, the use of fiscal policy for environmental protection, and the use of earmarked ecological funds for financing environmental remediation in Poland.

Piotr Gliński is Head of the Division of Civil Society of the Polish Academy of Sciences' Institute of Philosophy and Sociology in Warsaw. He has been instrumental in Poland's ongoing interdisciplinary study program on 'Man and Environment'. He has been affiliated with the World Future Studies Federation at the Interuniversity Centre, Dubrovnik, and has been a Ford Foundation scholar at Stanford University. Dr. Gliński is co-founder of the Social Ecological Institute and the Service Office for the Environmental Movement, an NGO dedicated to improving the management and organizational skills of environmental groups in Poland. He is the author of *Polish Greens: Social Movement in the Time of Transformation* (1996) and over forty articles on the sociology of culture and social movements. He holds a Ph.D from Warsaw University.

Jerzy Jendrośka is a member of the Environmental Law Research Group of the Polish Academy of Sciences' Institute of State and Law in Wrocław, Poland. He is co-founder and since 1987 Secretary of the Board of the Polish Environmental Law Association. Dr. Jendrośka has been instrumental in drafting new environmental legislation in Poland. And he has published numerous books and articles on environmental law, the institutional framework of compliance monitoring, environmental impact assessment procedures, public access to environmental information, and public participation in environmental decisionmaking. He currently serves as Poland's official representative to the United Nations Economic Commission for Europe's conference on the draft convention for public participation in environmental decisionmaking. He holds a Doctor of Law degree from Wrocław University.

Jerzy Śleszyński is a member of the Economics faculty and the Warsaw Ecological Economics Center at the University of Warsaw. He is among Poland's pre-eminent environmental economists, specializing in the use of economic instruments in ecological policy. He has published several articles and research papers on subjects including the use of economic instruments in Polish environmental policy and innovative financial mechanisms for nature conservation in Central and Eastern Europe. He holds a Ph.D in economics from Warsaw University.

Preface

In 1996, Indiana University sponsored a two-day conference on 'Environmental Protection in Poland and the Region'. This book is a product of that conference. Its purpose is to compensate for the lack of scholarly attention to environmental issues in transitional economies. This lack of attention is most perplexing given socialism's well-known environmental legacy of devastation.

In Poland and the rest of Central and Eastern Europe, environmental problems figured prominently in the demise of the old regime. And during the decade leading up to the fall of communism in 1989-91, Western journalists and scholars focused a great deal of attention on communist 'ecocide'. But since the beginning of the 'transitions' in 1989 western journalists and scholars have paid scant attention to environmental problems and policies. They have analyzed and re-analyzed political and economic changes in post-Communist countries, while all but ignoring the important ways in which those changes have influenced, and have been influenced by, environmental issues.

This is all the more disappointing because the countries of Central and Eastern Europe provide a unique opportunity to study the triadic relationship between political-economic system, environmental policies, and environmental performance. Such a study could serve to identify institutional prerequisites for (more) successful environmental protection, and not just for countries-in-transition. This book takes a first step toward that end.

The contributors to this volume address various aspects of Poland's environmental problems and policies during the period of systemic transformation. The single-country focus is designed to provide a common context for legal, economic and socio-political analyses. And Poland was an obvious choice because, before the fall of communism, it was reputed to be among the most polluted countries in the world, despite the fact that its environmental protection regime was the most sophisticated and well-established in the former Soviet Bloc. In addition, Poland was the first Soviet Bloc country to start down the transition path to market democracy. And, on several measures, it has travelled farther down that path than any other post-Communist country. Finally, information about the state of Poland's environment is relatively accessible and reliable.

This collection was conceived with the support of many individuals and institutions. Indiana University's School of Public and Environmental Affairs and Polish Studies Center co-sponsored the 1996 conference that led to this book. Tim Wiles, Director of the Polish Studies Center, co-organized the conference with Dan Cole. We are also grateful for the support of our respective home institutions, the Hudson Institute and the Indiana University School of Law at Indianapolis. Our greatest debt, however, is reserved for the contributors, who we consider friends as well as colleagues, for the high quality of their work and their collective patience.

John Clark and Daniel H. Cole
Indianapolis
May, 1998

1 Poland's Environmental Transformation: An Introduction

DANIEL H. COLE AND JOHN CLARK

This introductory chapter has two purposes: first, to present important background information about environmental protection in Poland before and during the transition to market democracy; and second, to provide a general analysis of the determinants of environmental performance, including Poland's environmental policies and the many institutions that have, at different times and in various ways, impeded or facilitated those policies. Along the way, we introduce certain themes and issues raised in subsequent chapters, which include contributions from some of the most prominent Polish and American economists, legal scholars, sociologists, and policy analysts writing about Poland today. Their varied expertise and perspectives, all focused on the single issue of environmental protection, make this book unique and a valuable contribution to our understanding of post-communist transitions.

A Tale of Two Polands

The Dirtiest Social Order Ever Constructed

Communism may have been the dirtiest social order ever constructed, and People's Poland was perhaps the dirtiest of communist countries. By the time Poland's Communist Party relinquished power in 1989, more than a third of all Poles lived in 'ecological danger zones' covering 11 per cent of the country's territory. Almost half of Poland's rivers and lakes were so polluted that their waters could not be used by industry, let alone consumed by humans. Polish air pollution levels were among the world's highest in absolute terms, per capita and per unit of GNP. And when wind and rain carried the air pollutants to earth, they contaminated vegetation and poisoned livestock — lead and cadmium levels in soil samples taken near Kraków were the highest recorded anywhere in the world (Cole, 1995a, p. 295).[1] Consequently, some 70 per cent of Poles in highly polluted regions, such as Kraków and Katowice, lived in health-threatening conditions. In the 1980s infant mortality in Katowice was 17.6 per 1000 live births, compared to 16.1 for Poland as a whole, 7.7 in France, and 5.8 in Sweden. And two-thirds of all children in the region suffered by age ten from 'serious mental and physical disabilities' (Cole, 1991, p. 215).

While pollution in Poland threatened public health, it also took its toll on the economy. During the 1980s cost estimates of environmental degradation and resource wastage ranged between 10 and 20 per cent of GNP (Ministry of Environmental Protection, 1992, pp. 40-3; Famielec et al., 1991, p. 25; Kabala, 1985, pp. 27, 30). These losses contributed to a decade-long economic downturn in Poland, which spelled the end of communism and, ironically, may have spared Poland even greater environmental degradation. The collapse of industrial output — especially in 1988 and 1989, when the Polish economy went into hyper-stagflationary free-fall — significantly reduced air and water pollution levels. In 1990, when gross domestic product fell 12 per cent and industrial production dropped by nearly a quarter, emissions levels of most major air pollutants declined between 17 and 23 per cent (see Table 1.1). The trend continued into 1991 when GDP and industrial output fell another 7 per cent and 12 per cent, respectively (see Table 1.1). This was not good news, of course, for the Polish economy or for millions of Polish workers who lost their jobs, but at least it made breathing a little easier.

Improvements In Transition

Some observers optimistically imagined that the collapse of communism created an opportunity for ecological rebirth in Poland and throughout Central and Eastern Europe. Roger Manser wrote that '[t]he recession threatened deindustrialization in some areas,' potentially creating 'a clean slate upon which to build an environmental nirvana'. All that stood between Poland and a greener future, according to Manser, were short-sighted politicians and profit-greedy businesses whose radar screens did not register environmental issues, and who thus wasted the opportunity for environmental renewal. In 1993 Manser noted disappointedly that '[f]rom an environmental point of view, virtually no fundamental changes had occurred in the way the economy was managed' (Manser, 1993, p. 148).

At least Manser began as an optimist. Others were pessimistic from the very start. On their view, the collapse of the communist political-economy and its replacement by unfettered capitalist markets could only make environmental matters worse. In 1990, when asked about the possibility of preventing environmental degradation while shifting to a market-based economy, a leading Hungarian economist replied, 'I do not think it is possible. . . . We have to accept the fact that private businessmen do not have the experience and they do not know how to think ahead for the long term' (Ivan Dobrovits, quoted in Levy et al., 1991, p. 20). Barbara Jancar-Webster (1995, p. 59), a noted author on environmental protection in socialist economies, wrote in a similar vein that '[a]ll predictions foresee a worsening of environmental conditions in the 1990s before the economic situation will have sufficiently improved so that governments can turn to the environment'. The implication: during the transition, economic and environmental issues were not only separate but inherently in conflict.

Many environmental groups were equally pessimistic, though for somewhat different reasons. To them, the transition from command economy and Leninist autocracy to private capital and democratic governments represented no fundamental change. The real environmental culprit was industrialization, whether of the Soviet or the American variety (see Gliński, 1996, pp. 209-11). The environmental pessimism was so pervasive and virulent that, for a while at least, even cold hard facts could not overcome it. The pessimists explained away evident environmental improvements as mere transitory symptoms of economic recession, and predicted that pollution levels would grow higher than ever once the recession ended and economic growth resumed (see, e.g., Manser, 1993, p. 75). But they were wrong.

Poland's economy emerged from recession in 1992; its GDP and industrial production figures have been rising ever since. Since 1994 Poland has maintained one of the highest economic growth rates in all of Europe. Meanwhile, despite the pessimists' predictions that pollution levels would rise along with the Polish economy, most important measures of pollution have continued to decline significantly since 1992 (see Table 1.1 below).

No one predicted anything close to the pollution reductions witnessed during and after Poland's economic recession. Indeed, when Poland's Environmental Protection Ministry promulgated a new National Environmental Policy (NEP) in 1990 (Ministry of Environmental Protection, Natural Resources and Forestry 1991), its 10-, 15-, and 20-year targets seemed over-optimistic. Yet by 1995 Poland already had met or exceeded goals established for the year 2000 for emissions of sulfur dioxide, dust (particulates), nitrogen oxides, and other air pollutants (compare Table 1.1 below and Ministry of

Environmental Protection, Natural Resources and Forestry, 1991). This suggests that something fundamental has changed for environmental protection since the collapse of communism in Poland.

Table 1.1 Percentage Changes in GDP, Industrial Production, and Air Pollution Emissions in Poland, 1990-5

The economy fell and rose again -

	GDP	Industrial Production	Dust	Carbon oxides	Nitrogen oxides	Sulfur oxides
1990	-12	-24	-23	-17	-18	-21
1991	-7	-12	-21	-28	-6	-8
1992	2	4	-26	-15	-8	-10
1993	4	7	-13	-21	0	-3
1994	5	13	-12	0	0	-3
1995	7	9	-18	-12	1	-5

- but pollution emissions just kept falling

Source: Cole (1997, p. 192)

Assessing Environmental Protection in Post-communist Poland

This book combines a surfeit of detailed information about the state of environmental protection in post-communist Poland with the perspectives of several different academic disciplines. The various chapters contribute in different ways to our understanding of Poland's ongoing political-economic transformation and how the institutional and policy changes accompanying that transformation affect environmental protection. And yet, the different perspectives yield a fairly clear and consistent picture of Poland's environmental problems and policies. Indeed, read together, the chapters of this book together present a unified assessment by a *community* of nongovernment experts in Polish environmental policy.

Each of the contributors brings unique credentials to the table. Jerzy Śleszyński and Bogusław Fiedor are among the top environmental economists in Central Europe. Jerzy Jendrośka is a leading environmental law expert, who has played a major part in drafting environmental legislation in post-Communist Poland. Sociologist Piotr Gliński has been an analyst and sympathetic critic of the Polish environmental movement since the early 1980s. Even the American contributors to this book belong to Poland's environmental policy community. Through several projects at the Hubert H. Humphrey Institute of Public Affairs at the University of Minnesota (many of which have been funded by the US Agency for International Development), Zbigniew Bochniarz and Richard Bolan have helped shape Poland's environmental policy agenda by training a new generation of Polish environmental policy experts and by cultivating a more environmentally-conscious business sector in Poland. Attorney Susan Cummings went to Poland as a US-funded advisor to the Ministry of Privatization, where she played a pivotal role in designing institutions to deal with environmental issues in privatization transactions. She then joined one of Poland's National Investment Funds in Poland's ambitious Mass Privatization Program. Her contribution to this book provides the valuable perspective of the regulated industries.

Somewhat surprisingly, this diverse collection of scholars agrees on almost all major issues relating to environmental policy. They agree that government has a vital role to play in environmental protection, though its ability to enforce environmental regulations is clearly limited. Thus, market mechanisms for environmental protection are also necessary. All agree that the government should use, rather than simply limit, the market. Market-based incentives, such as effluent taxes, should complement regulatory controls to ensure that the polluters pay for the costs they impose on society. Subsidies, in the form of low-interest loans, may also provide appropriate incentives for private producers to make environmentally-friendly investments. This broad consensus within the environmental policy community has informed most of Poland's environmental policy changes since 1989.

But Poland's improvements in environmental protection are not only, or even predominantly, the result of policy changes.

The Old System's Failures Laid the Groundwork for the New System's Successes

Environmental policies invariably operate within an overarching political-economic system, the institutions of which may largely determine the success of failure of environmental protection efforts (see generally Cole, 1998). Indeed, the most surprising aspect of Poland's remarkable environmental improvements may be how little formal environmental policy — the laws and regulations on the books — has actually changed since 1989. There have been a number of important policy innovations, many of which are discussed in the pages of this book, but, generally speaking, Poland's environmental protection regime remains much as it was under communism. As Jerzy Jendrośka notes in Chapter 5, Poland's 1980 Environmental Protection and Development Act is still largely intact, although it has undergone a series of (mostly minor) amendments since 1989. As Richard Bolan and Zbigniew Bochniarz explain in Chapter 2 and Jerzy Śleszyński further explains in Chapter 3, Poland continues to rely heavily on a system of environmental taxes that has been in place since 1980; the rates have gone up since 1989 but the structure of the environmental tax system remains the same. There is, in fact, great continuity in Poland's system of environmental protection, dating all the way back to the pre-War (i.e., pre-Communist) era (see Cole, 1995c).

Poland's remarkable environmental improvements have been facilitated by important institutional developments since 1989. First and foremost, Poland became a democracy committed to the rule of law. Regardless of what one thinks of the former communists who won control of the Polish Parliament in 1993 and ousted Lech Wałęsa from the presidency in 1995, or about the post-Solidarity alliance that wrested power back from the post-communists in 1997, what matters most is that political power was transferred peacefully, orderly, and according to law. 'What was remarkable about the elections was that they were so unremarkable,' noted Timothy Garton Ash in 1997. 'Poles acted as though they had been voting in free elections for decades' (Ash, 1997, 9.C1).

Second, Poland possesses the strongest market economy in the post-communist world. Ironically, the first set of market reforms were instituted by Poland's last Communist government of Prime Minister Mieczysław Rakowski, in a burst of free-for-all *laissez-faire*, a full year before Leszek Balcerowicz's 'shock therapy' reform program took effect. Today Poland's private sector is the driving force behind a burst of strong economic growth that has lasted five years and shows no signs of slowing. Third, Poland has an active and engaged civil society, a dense web of formal and informal groups operating outside the sphere of the state. Unlike elsewhere in the former Soviet bloc, this civil society did not emerge as a wholly new phenomenon with the demise of communism. Indeed, the diverse and self-assertive civil society of People's Poland — exemplified but by no means limited to the trade union and social movement known as 'Solidarity' — deserves a large measure of credit for the destruction of the old system.

The essays in this book examine the generally beneficial implications for environmental protection of democracy, markets and civil society in Poland. Several of them suggest that Poland's post-communist environmental successes are best seen as unintended consequences of disastrously unsuccessful communist-era policies. The failed efforts of the communist era laid the groundwork for the more successful efforts of the immediate post-communist transition period. The environmental policies adopted by the authoritarian communist regime, as it clung to a woefully unproductive command economy and repressed an assertive anti-communist opposition, all failed badly. But the formal policies were not nearly as defective as the system in which they operated. Once the system was transformed — markets were freed from central control, democratic institutions were adopted, the rule of law was instituted — the very same communist era policies began to provide more effective environmental protection.

We have no shortage of explanations for the ecological failings of the communist political-economic system, ranging from the liberal (Cole, 1998) to the libertarian (Hill, 1992) to the post-Marxist (Bahro, 1984). In Chapter 4 of this volume, Bogusław Fiedor (following Kornai, 1992) provides an explanation based on endemic soft budget constraints in the socialist economic system. But less well explained are two separate reactions to Poland's ecological crisis. For entirely different reasons and largely unintentionally, the Communists and the opposition acted in ways that have shaped the course of environmental protection during the post-communist transition period.

As a matter of ideology or expediency, the Communist Party routinely denied the very existence of many social problems. But during the 1970s and 1980s the Party grew increasingly open about Poland's environmental problems. It would have been futile, of course, for Party leaders to deny the undrinkable water and unbreathable air in several regions of Poland. Moreover, for all their privileges, the Party officials themselves could not easily avoid the effects of living in a polluted environment.[2] So, during the 1970s, they named environmental protection a top political priority (see Cole, 1998, pp. 38, 41). And in 1980, just months before the Solidarity movement was to completely redefine the nature of Polish politics, the authorities enacted the Environmental Protection and Development Act (EPDA) (1980 Dz.U., No. 3, item 6).

As Jerzy Jendrośka and others discuss in this volume, the EPDA was the most market-oriented environmental statute adopted by any communist country; indeed, it may have been the most market-oriented measure adopted by any country in the world up to that point. Well in advance of the current worldwide trend toward replacing expensive 'command-and-control' regulations with market-based incentives,[3] Poland's 1980 EPDA introduced the world's most comprehensive system of resource and pollution taxes, as well as noncompliance fines (see Cole, 1998, pp. 70-1). But there was a snag. Market-based incentives require markets to be effective (Cole, 1998, p. 71). The socialist economic system lacked the kind of rational pricing system, competitive markets and hard budget constraints necessary for a tax-based approach to work. Consequently, as pollution-control measures, the market-mechanisms in Poland's 1980 EPDA were a miserable failure.

The Party's misguided efforts to cure non-market environmental problems with market-based solutions was not the only important response to Poland's environmental problems to emerge in 1980. As Piotr Gliński details in Chapter 7 of this volume, that year also witnessed the birth, along with Solidarity, of Poland's independent environmental movement.[4] Like all other aspects independent civil society that erupted during Solidarity's brief first life, much of the independent environmental movement was driven underground when Martial Law was declared on 13 December 1981. Certain segments of the movement, however, such as the extensive network of Polish Ecology Clubs, were legally permitted to continue operating. This limited tolerance of environmental groups proved to be important for the continued development of environmental protection in communist and post-communist Poland because, under the 1980 EPDA, legally recognized environmental organizations, such as the Polish Ecology Club, were granted rights, broad even by West European standards, to participate in litigation and administrative hearings. Of course during the Communist era these rights did not amount to much. Thanks to the system's fundamental and unreformable investment 'fetish' and the priority communism accorded to producers — particularly to the largest (and dirtiest) producers — most suits by environmental groups were dismissed because of the 'higher needs' of the state (see Cole, 1998, pp. 81, 103).

Among the most important unintended effects of citizen participation rights during the 1980s was the formation of what Jerzy Jendrośka in Chapter 5 (p. 87) calls 'a basic infrastructure for developing a successful environmental policy' in Poland. Jendrośka describes an intriguing and important civil society component of this infrastructure: the development during the 1980s of a network of more than a hundred lawyers and legal experts specializing in environmental law. Most had been trained or conducted research in two of the oldest and most important environmental law centers in the communist world — at Warsaw University and the Polish Academy of Sciences' Environmental

Law Research Group in Wrocław. Some of these environmental law specialists worked for industrial ministries, in effect as in-house lawyers, on environmental litigation or voluntary environmental damage compensation packages. Others provided *pro bono* services to the environmental NGOs that were legally authorized to file legal claims against polluters. Many lobbied unsuccessfully, from within or without the Party, for future reforms. Most came to view themselves as members of a fraternity, whose legitimate and proper professional development was hampered by the country's irrational legal order. Their fraternity and professionalism was accepted by the Communist regime, which officially registered the Polish Environmental Law Association in 1986 as a legally recognized, independent NGO. This Association grew to become a vital part of Poland's post-communist environmental policy community.

The development of environmental law expertise in Poland bore fruit in April of 1989, during the 'Round Table' talks between the Communist government of Mieczysław Rakowski and the Solidarity opposition, which marked the beginning of the end of communism in Poland. One of the 'Round Table's' 14 subgroups was charged with defining the needs and orientation for a new environmental policy. As Jendrośka mentions in Chapter 5 (p. 87), members of the Polish Environmental Law Association sat on both sides of the negotiating table. And there was substantial consensus among them that the current system was badly in need of change. This made the ecological subgroup one of the most cooperative and least contentious subgroups of the 'Round Table'. The result was an Ecological Protocol that specified (among other mechanisms) several provisions guaranteeing the rights of civil society to participate in environmental decisionmaking, including the development of future legislation concerning freedom of access to environmental information, the right to conduct and freely publish environmental research, the right to sue to protect the environment, and the exclusion of environmental information from state secret laws (see Cole, 1995b, p. 285).

As already noted, the various communist-era environmental reforms were all miserable failures so long as the system persisted. That is not surprising, given the nature of the system and the design of the reforms. The history of Communist Poland is littered with a trail of half-hearted and half-baked decentralizing economic reforms, usually adopted in response to a worker uprising or economic collapse, doomed to be forgotten as soon as the Party/state reasserted its control over society and economy. One might think the same of environmental policies left over from the communist era, which is why it is surprising that these communist-era environmental reforms contributed substantially to a cleaner Poland after communism.

State, Market or Civil Society: Who is Cleaning Up in Poland?

Several institutional factors have contributed to the declining rate of pollution in Poland over the last several years. Consider first the establishment of democracy and the rule of law. As many of the authors in this volume address (either directly or indirectly), the new order in Poland is one in which the law matters. Adopted one decade and a social order ahead of its time, the 1980 Environmental Protection and Development Act was based on principles that have since been widely embraced by Western environmental law experts. In post-communist Poland the Act has been amended but not replaced.[5] Today it operates in a genuine market environment, in which hard budget constraints predominate and the government no longer faces a regulatory conflict of interest as both owner and regulator of polluters. Consequently, the Act's fees and fines have begun to effect the incentives of producers, inducing them to conserve resource inputs and reduce pollution. Moreover, as Zbigniew Bochniarz and Richard Bolan note in Chapter 2 (p. 21), Poland's environmental fees and fines today are among the world's highest.

One indication of the effectiveness of environmental protection in the new order is that powerful economic actors are now working hard (though, for the most part, unsuccessfully) to avoid the law's bite. In 1992, for instance, after the Ministry of Environmental Protection announced sizeable fee increases for air pollution emissions, many state-owned giants used their political clout to convince the Ministry to roll-back the charges. Environmentalists and privatized firms protested against the fee reductions. Even some state-owned enterprises that *had* invested in environmental improvements complained that the fee reductions devalued their investments. Ultimately, the Polish Council of Ministers reversed the Environment Ministry's decision and reinstated the higher fees (Cole, 1998, pp. 200-01). There are two morals to this story. First, democracy means transparency; the dismantling of the Leninist Party/state has meant that regulations are drafted, discussed, and implemented in the open. Second, the institution of a competitive market economy has served to disaggregate the interests of economic producers; no longer do industries present a united front for or against environmental regulations (see Cole, 1998, p. 201).

The regulatory conflict of interest that suffused the communist political economy — the state was both regulator and owner of the means of production— has been resolved for Poland's booming private sector and even for a large portion of the economy that remains in the state's hands. The 1991 Law on the State Environmental Protection Inspectorate (1991 Dz.U. No. 77, item 335) created an environmental law enforcement agency which, according to Jerzy Jendrośka in Chapter 5 (p. 97), has the power 'to impose non-compliance fines, to halt activities endangering the environment, and to ban the sale or import of raw materials and other products that fail to meet Polish environmental standards'. Under this law, the Inspectorate has been a very effective environmental watch-dog, so effective in fact that large state-owned producers and the regional political authorities that depend on them for jobs and tax revenue have tried repeatedly and without success to close down the Inspectorate (see *Wprost*, 18 June 1995).

The establishment of democratic institutions is only part of the story of Poland's transformation over the past decade, and only part of the cause of falling pollution rates. Perhaps even more important have been the effects of market institutions. Most economists agree that the combination of competitive markets, scarcity pricing and hard budget constraints encourage resource conservation in production. Resource prices that accurately reflect the rising costs of development and depletion will induce industrial users to improve dynamic efficiency in production. Over time, users will find it cost effective to develop technologies and processes to obtain more production from each input resource unit. The result is less resource waste and pollution per unit of production. This trend was not evident in Communist Poland or other states with socialist economic systems because they lacked competitive markets, scarcity pricing and hard budget constraints. But it has already been evident in post-communist Poland, where, between 1992 and 1994, the energy intensity of GDP dropped by 13.6 per cent (Cole, 1998, p. 212). Meanwhile, Poles are discovering that 'green business' is big business — a billion-dollar-a-year industry that is expected to expand several-fold in coming years (see *Financial Times*, 28 February 1996). As Zbigniew Bochniarz and Richard Bolan explain in Chapter 2 of this volume (p. 26), the market offers incentives to firms that specialize in cleaning up the environment, with more than 1,500 Polish firms and more than 100 foreign companies participating in this booming market.

The third set of changes in Poland contributing to falling pollution levels relates to civil society. Since the fall of communism, as Piotr Gliński notes in Chapter 7, hundreds of locally- and nationally-based environmental NGOs have mushroomed across Poland. Some have contributed to the reduction of pollution by lobbying the government for stricter environmental laws. Others monitor the behavior of both regulators and regulated industries, publicizing lapses in compliance and enforcement. Still others seek to educate the public and economic producers about environmental problems, policies and practices. Despite wide variations in their orientations, goals and methods, Poland's environmental groups have networked effectively with one another and with such social organizations as labor organizations and political parties. The Independent Environmental Team, a group of environmental experts who left government service when the former Communists returned to power in 1993, prepares environmental analyses and policy recommendations for the Solidarity union and opposition parties (see *Tygodnik Solidarność*, 31 March 1995). In some cases, such as the dispute over environmental fee increases discussed earlier, environmental organizations have successfully allied with business interests.

But Do All Good Things Necessarily Go Together?

It is clear that, so far at least, Poland's various political and economic reforms have almost automatically facilitated improved environmental performance. Government regulation, market competition, and NGO activities have not conflicted with and, in some cases, have actually reinforced one another. Some environmental organizations have discovered that they can be more effective when the work with, rather than against, business and political groups. More importantly, Polish environmental law experts have learned (from the disastrous performance of the EPDA during the 1980s) that laws do not operate in political, social or economic vacuums; market-based regulations in particular require functional markets to be effective.

The symbiotic relation between markets and economic regulation is reflected in the uses to which collected environmental fees and fines are put. In order to protect the environmental protection budget from political manipulation and to insulate it from economic downturns, the Polish government dedicates collected environmental fees and fines to an innovative set of local, regional and national environmental funds, which are discussed by Bochniarz and Bolan in Chapter 2, by Śleszyński in Chapter 3, and by Jendrośka in Chapter 5. The funds are managed by autonomous Boards of Directors representing environmental NGOs, local and regional authorities, and the Ministry of Environmental Protection, Natural Resource and Forestry. They provide resources for national environmental projects, such as sewage treatment plant construction, and low-interest loans for private environmental investments. Relying more heavily on loans rather than outright grants (subsidies) for private environmental projects forces fund administrators and borrowers to calculate capital needs and expected revenues carefully to ensure that (1) environmental investments are efficient, and (2) funds will revolve for future environmental investment opportunities. In short, the funds represent an innovative means by which political authorities can harness market forces for socially adopted goals.

The fact that the market, the state, and civil society have reinforced one another has been good news for Poland's natural environmental. But much remains to be done, both to clean up the Polish environment — still one of Europe's most polluted — and to complete the country's broader political, economic and social transformation. Despite the booming private sector, for instance, most of Poland's dirtiest heavy industrial enterprises are still owned by the state. Further exposing these giants to market forces (e.g., through accelerated privatization) would help immensely to further the clean-up. Large chunks of the country's legal and regulatory infrastructure remain under construction. And nongovernmental environmental groups acknowledge that they still have a lot to learn about participating as professionals in environmental policy making.

From the 'Easy' to a More 'Interesting' Stage of Environmental Protection in Poland

There are good reasons for supposing that Poland's environmental improvements will continue into the future. For one thing, Poland still has a long way to go to achieve West European standards of environmental health and safety. As several authors in this volume mention (if only in passing), Poland's virtual national obsession with joining western political, economic, and security institutions, including the OECD, NATO, and most importantly, the European Union, creates a strong incentive for continued environmental progress. To gain admittance, Poland will have to do more than simply 'harmonize' its environmental laws with EU Directives. As Cole (1997, p. 211) has observed, '[m]ere paper changes unaccompanied by implementation and enforcement efforts are unlikely to impress current EU members. Poland will have to show continued *actual* progress in environmental protection'. This should not be a problem so long as state, market and social institutions continue to work smoothly together for environmental protection. But how likely is it that they will?

As noted earlier, Poland's environmental improvements have, so far, been virtually automatic consequences of systemic reforms, most notably the introduction of harder budget constraints and the rule of law (see Cole, 1998, especially Chapter 8). In this respect, Poland's experience generally supports claims of a positive correlation between environmental performance and democratic capitalism (see, e.g., Payne, 1995 and Lewis, 1992). Certainly, Poland's embrace of democratic institutions, markets, civil society, and the international community has brought substantial environmental rewards. Laws and regulations are now enforced that were un- or under-enforced during the communist era. Post-communist Polish governments have responded to democratically articulated pressures to clean up pollution and protect natural resources. Market forces have driven some of the dirtiest enterprises out of business and have forced many others to reduce their waste. Nongovernmental environmental organizations have, in many instances, influenced public policy. And Poland's efforts to join the West have led to it adopt cleaner standards. Poland's dramatic environmental improvements are no fluke; they are directly related to the replacement of communist institutions with those of market democracy.

It has not been simply a matter of institutional change, of course. As made clear in this and the other chapters of this volume, policy innovations in post-communist Poland, including novel funding mechanisms such as the Environmental Protection Bank and stronger enforcement authorities, certainly have played a big part in Poland's environmental improvements. But, in the absence of the changes to state, economic and social structures, those policy innovations would likely have been completely ineffective. The communist system simply lacked the markets, the commitment to the rule of law, and the respect for civil society that successful environmental policies appear to require. Conversely, even if Poland had not innovated any new policies in the post-communist era, its systemic reforms, by themselves, would likely have generated substantial pollution reductions.

Thus, from an environmental point of view, it may be good news that Poland's systemic reforms are far from complete. As of late 1997, some 3,500 enterprises have yet to be privatized (or even 'commercialized'), and entire sectors of the economy, including agriculture, banking and mining, still await restructuring. Consequently, further (relatively) cheap environmental improvements are available to be exploited. Poland's economy has a tremendous capacity for additional efficiency gains that would conserve resource inputs and reduce pollution emissions per unit of production. Budget and law constraints remain relatively soft in many industries still under state ownership, including, most notably, the mining industry. And the Polish government has only just begun to regulate, in a systematic way, certain forms of pollution such as hazardous wastes.

The bad news is that the pace of systemic reform has slowed since 1992. Not coincidentally, there are signs that the 'easy' stage of virtually automatic environmental improvements in Poland is being replaced by what may prove to be an even more 'interesting' stage.

According to a preliminary report by the State Environmental Protection Inspectorate, aggregate air pollution in Poland increased slightly in 1996 (see *International Environment Reporter — Current Report*, 9 July 1997).[6] This increase could be an artifact of an unusually cold winter in 1996, as the Chief Inspector has suggested. Nevertheless, it marks an ominous break from a consistent, decade-long trend of emissions reductions.[7]

If the 'easy' stage of environmental improvements in Poland was characterized primarily by institutionally-driven efficiency gains and enhanced environmental law enforcement, this new, more 'interesting' stage may be characterized by reduced social consensus on environmental goals, more difficult policy choices requiring environmental and economic trade-offs, more varied and contentious interest-group politics, heightened social conflict and discord, and quite possibly public misunderstanding. Continued improvements will not come cheaply: in 1997, officials from the Polish Ministry of the Environment and the World Bank estimated that Poland would have to spend between 3.5 and 4.6 per cent of its GDP on environmental projects over the coming years — more than double the current level of environmental spending — in order to meet the binding standards of the European Union (*Polish News Bulletin*, 15 October 1997). One thing seems certain: environmental improvements during this more 'interesting' stage will be anything but automatic. They will depend increasingly on Poland's ability to implement and enforce sound policies.

Notes

1. For more complete accounts of Poland's environmental crisis under socialism see Cole (1998), Cole (1991), Nowicki (1993), and Ministry of Environmental Protection, Natural Resources and Forestry (1992).
2. Even during the heady days of privileged junketing during the Gierek Administration in the early 1970s, the nomenklatura had to return to live in their polluted homeland every now and then. For an analysis of the excesses of this period and how the rulers tried to cope with the consequences, see Clark and Wildavsky (1990).
3. For an overview of this trend, see Barde and Opschoor (1994).
4. Gliński (1996) presents a more elaborate treatment of the birth and evolution of Poland's environmental movement. The best English-language source is Hicks (1996).
5. The Polish Environmental Law Association recently drafted legislation that, if enacted, would completely replace the EPDA. However, its prospects for passage in the current *Sejm* are poor.
6. Poland's Main Statistical Office will release final, official environmental statistics for 1996 in January or February of 1998. These preliminary statistics are based on reports from the State Environmental Protection Inspectorate.
7. It is worth noting that, according to the State Environmental Protection Inspectorate's report, other environmental indicators continued to improve in 1996 (*International Environmental Reporter — Current Reports*, 9 July 1997).

References

1980 Dziennik Ustaw (hereinafter 'Dz.U.') No. 3, item 6, *Ustawa z dnia 31 stycznia 1980 r. o ochronie I kształowaniu środowiska* (Law of 31 January 1980 on environmental protection and development).

1991 Dz.U. No. 77, item 335, *Ustawa z dnia 20 lipca 1991 r. o Państwowe Inspekcji Ochrony Środowiska* (Law of 20 July 1991 concerning the State Environmental Protection Inspectorate).

Ash, Timothy Garton (1977), 'Eastern Europe's Paradox: Why Some Nations Prosper While Others Decline,' *The Washington Post* (5 October), p. C1.

Bahro, Rudolf (1984), *From Red to Green: Interviews with New Left Review*, Verso, London.

Barde, Jean-Phillippe and Johanes Baptist Opschoor (1994), 'From Stick to Carrot in the Environment', *OECD Observer*, no. 186 (February/March), pp. 23-7.

Clark, John and Aaron Wildavsky (1990), *The Moral Collapse of Communism: Poland As a Cautionary Tale,* Institute of Contemporary Studies, San Francisco.

Cole, Daniel H. (1998), *Instituting Environmental Protection: From Red to Green in Poland,* Macmillan, Basingstoke (UK) and St. Martin's, New York.

Cole, Daniel H. (1995a), 'Environmental Protection and Economic Growth: Lessons from Socialist Europe', in R.P. Malloy and C.K. Braun (eds), *Law and Economics: New and Critical Perspectives*, Peter Lang, New York, pp. 295-329.

Cole, Daniel H. (1995b), 'Poland's Progress: Environmental Protection in a Period of Transition', *The Parker School Journal of East European Law*, vol. 2, pp. 279-319.

Cole, Daniel H. (1995c), 'An Outline History of Environmental Law and Administration in Poland', *Hastings International and Comparative Law Review*, vol. 18, pp. 297-358.

Cole, Daniel H. (1991), 'Cleaning Up Kraków: Poland's Ecological Crisis and the Political Economy of International Environmental Assistance', *Colorado Journal of International Environmental Law and Policy*, vol. 2, pp. 205-45.

Famielec, Józefa, Kazimierz Górka and Grażyna Mojżesz (1991), 'Economic Losses Due to Environmental Pollution in Poland', in K. Górka (ed), *Environmental and Economic Aspects of the Industrial Development in Poland*, Cracow Academy of Economics, Kraków, pp. 43-53.

Gliński, Piotr (1996), *Polscy Zieloni. Ruch społeczny w okresie przemian*, Wydawnictwo IFiS PAN, Warsaw.

Hicks, Barbara (1996), *Environmental Politics in Poland: A Social Movement Between Regime and Opposition*. New York: Columbia University Press.

Hill, Peter J. (1992), 'Environmental Problems Under Socialism', *Cato Journal* (Fall), pp. 321-35.

Jancar-Webster, Barbara (1995), 'Environmental Degradation and Regional Instability in Central Europe', in J. DeBardeleben and J. Hannigan (eds), *Environmental Security and Quality after Communism: Eastern Europe and the Soviet Successor States*. Boulder: Westview, pp. 43-68.

Kabala, Stanely (1985), 'Poland: Facing the Hidden Cost of Development', *Environment* (27 November), pp. 6-42.

Kornai, János (1992), *The Socialist System: The Political Economy of Communism*. Princeton: Princeton University Press.

Levy, Barry S., Charles Levenstein, and Eve Sprangler (1991), 'Environment and Health in Eastern Europe', *New Solutions*, vol. 1 (Winter), pp. 18-24.

Lewis, Martin (1992) *Green Delusions: An Environmentalist Critique of Radical Environmentalism*. Durham: Duke University Press.

Manser, Roger (1993), *Failed Transitions: The Eastern European Economy and Environment Since the Fall of Communism.* New York: The New Press.

Ministry of Environmental Protection, Natural Resources and Forestry (1992), *Environment in Poland: Issues and Solutions.* Ministry of Environmental Protection, Natural Resources and Forestry, Warsaw.

Ministry of Environmental Protection, Natural Resources and Forestry (1991), *National Environmental Policy of Poland.* Ministry of Environmental Protection, Natural Resources and Forestry, Warsaw.

Nowicki, Maciej (1993), *Environment in Poland.* Kluwer, Dordrecht.

Payne, Rodger A. (1995), 'Freedom and the Environment', *Journal of Democracy*, vol. 6(3), pp. 41-55.

Timberlake, Lloyd (1981), 'Poland — the most polluted country in the world?', *New Scientist*, vol. 92, pp. 248-50.

2 Sustainable Institutional Design in Poland: Putting Environmental Protection on a Self-Financing Basis

ZBIGNIEW BOCHNIARZ AND RICHARD S. BOLAN

Introduction

Socialist central planning left Poland with not only a devastated economy subject to hyperinflation and a huge foreign debt, but also a heavily damaged environment. In order to tackle these problems Polish leaders chose to implement radical reform programs often referred to as 'big bang' or 'shock therapy'. Public trust in Poland's first democratically elected government since World War II allowed this ambitious but painful program to be implemented, starting in January 1990.

The market-oriented approach of the Polish reform program impacted environmental protection by putting environmental protection on a self-financing basis. The institutional vacuum left after the collapse of the old system was filled by a new institutional framework adapting Poland's environmental protection system to market conditions. In effect, Poland has sought to create an off-budget market for environmental protection in order to avoid political pressures to cut or reduce state budget expenditures on environmental protection. The system is also designed to create revolving funds that can provide a sustainable base for ecological improvements. In addition to a more predictable revenue base, this framework allows better control over the effectiveness of policy implementation.

The framework of financing environmental protection and sustainable development projects is based on the following institutions: (1) the National Fund for Environmental Protection and Management; (2) 49 environmental and water management funds at the voivodship (regional administration) level; (3) 2460 *gmina* (local) environmental and water management funds; (4) the Environmental Protection Bank; (5) the Forestry Fund; (6) the Farmland Protection Fund; (7) funds owned by private and public businesses; (8) local and municipal budgets; (9) the central state budget; and (10) the EcoFund and other forms of foreign assistance.

The first three of these listed funding sources were based on the principle of self-financing environmental protection through the collection of fees and fines levied on polluters and users of environmental resources. In 1994 these environmental charges supplied about 41 per cent of all moneys expended on environmental protection in Poland. Of that 41 per cent, the largest portion — about 22 per cent — came from the National Fund for Environmental Protection and Water Management (NFEPWM); about 15 per cent came from voivodship environmental funds (VEPWMF); and the rest (4 per cent) came from the local environmental protection and water management funds (LEPWMF). The remaining 59 per cent of Poland's environmental protection budget came from four other sources: 31 per cent from the business sector (through environmental charges and interest payments on loans taken from the Environmental Protection Bank); 19 per cent from local and municipal budgets; 5 per cent from Poland's central budget; and 4 per cent from foreign assistance (GUS, 1995a, p. 361).

These figures show that Poland has developed a unique solution to the difficulties facing environmental protection by institutionalizing self-financing and decentralizing the funds. Taking this into account, this paper will address the following questions: what are the advantages and disadvantages of such solutions, what institutional arrangements are behind this framework, and what are the ecological and economic consequences of such a system?

Institutionalizing the Polluter Pays Principle

Poland's self-financing system for the environment is based on a basic precept of environmental economics: the 'polluter pays principle'.[1] Virtually every human activity has the potential to create environmental damage by exploiting resources or producing wastes. The 'polluter pays principle' requires all users of the environment to pick up the tab for their resource exploitation or waste production. One of the key questions, of course, has been how much should the user pay. Traditionally, it has been difficult to price non-commodity natural resources.

Few nations have been able to create a political and economic system conducive to heavy reliance on publicly-imposed environmental use charges. Poland, however, has tried to do so during its radical transition to market democracy since 1989.[2] It therefore represents an experiment in the operational demands of economy-wide resource use and effluent taxation.

The elaborate institutional framework that Poland has set up to implement this theory relies on market-type institutions for its administration. A basic fee structure is established by the Ministry of Environmental Protection, Natural Resources and Forestry (MEPNRF). Fees are charged for all uses of the physical environment: taking resources from the environment or emitting wastes into the environment. Specifically, users of the environment pay for consuming water, discharging sewage into waters or soils, emitting air pollutants, disposing of waste, removing trees and bushes, and extracting minerals.

For water use and other resource extraction or exploitation activities, fees are based on the extent of usage. For polluting and waste disposal activities, permits are required and charges are set on a graduated scale based on performance standards, thus creating built-in incentives for enterprises to invest in more efficient and cleaner technologies. For example, the fee for emitting one ton of sulphur dioxide (SO_2) or nitrogen oxide (NO_x) in 1994 was about $80 (US); emissions fees for carcinogenic air pollutants, such as asbestos, chromium, or benzene, exceed $51,000 (US) (1994 Dziennik Ustaw, No. 140, item 172). These, and Poland's other environmental fees, are among the world's highest.

The first significant rate increases became effective on 1 July 1990, and additional increases were instituted on 1 January 1992. Environmental charges temporarily were reduced during the fall of 1992, which caused the environmental funds to deteriorate by about 30 per cent; but higher rates finally were restored in April 1993 (Żylicz, 1995, pp. 203-18).

The success of Poland's self-financing system is due, in part, to monitoring and enforcement operations. At the national level, the State Inspectorate for Environmental Protection is charged with monitoring emissions and enforcing environmental standards. It operates as a central enforcement agency with regional divisions — the voivodship inspectorates.[3] These agencies are responsible for assuring that polluters and resource users have proper permits and comply with all permit conditions. Enterprises that pollute without a permit or fail to comply with permit conditions are subject to fines. Those fines, like the effluent fees, have been increased significantly during the past five years to insure that environmental users have a greater economic incentive to comply with, rather than violate, environmental norms. Like the effluent and resource use fees, collected noncompliance fines provide revenues for Poland's various environmental funds. But while the fines have increased markedly since 1989, they contribute only about three per cent to fund revenues. This is partly because of the high transaction costs involved in collecting environmental fines; the collection rate is quite low at about 13 per cent (GUS, 1995b, p. 46). For this same reason, fines have not been very effective as a policy instrument.

Environmental law enforcement in Poland has improved greatly since the early 1990s. Today the State Environmental Protection Inspectorate (SEPI) employs 3,000 inspectors operating out of 50 offices throughout Poland. Inspectors can shut down polluters, order installation of pollution-control equipment, and impose noncompliance fines. The increased power and staffing of the State Environmental Protection Inspectorate has led directly to improved monitoring, compliance, and enforcement (see Cole, 1995, pp. 308-11).

The Ecological Funds

The three main ecological funds in Poland — respectively, the National Environmental Protection and Water Management Fund (NEPWMF), the Voivodship (regional) Environmental Protection and Water Management Funds (VEPWMF), and the Local (*gmina*) Environmental Protection and Water Management Funds (LEPWMF) — are only indirectly subordinated to environmental authorities. They play the primary role in financing environmental investments, contributing over 41 per cent of all resources for environmental protection. Pollution and resource use charges account for nearly 97 per cent of their revenues (GUS, 1995b, p. 46). The funds split the collected charges: the National Fund receives 35 per cent of all collected fees; the Voivodship funds keep 55 per cent of all fees collected in their respective regions, and the local funds keep 10 per cent of all locally collected fees.[4]

The separation of the NEPWMF and the VEPWMFs from direct management by the environmental authorities represents a major institutional innovation. The funds are managed autonomously and supervised by a board of directors representing major stakeholders, including the Environment Ministry, regional and local government authorities, and representatives of non-governmental organizations (NGOs). The funds must, however, follow the policy guidance and priorities set by the Ministry and regional authorities. One major policy guidance requires that the funds distribute investment monies by loans rather than grants. Aside from educational, monitoring and some other special projects that utilize about 10 per cent of fund revenues, the balance is distributed through low-interest loans; when market interest rates stood at 30 per cent, the environmental funds were making loans at between eight and 12 per cent. The use of loans rather than grants has two major advantages. First, it significantly increases the efficiency of environmental investments by inducing potential borrowers to calculate carefully their capital needs and expected revenues. Second, the use of loans rather than grants ensures revolving funds; fund monies can be invested more than once.

The National Fund for Environmental Protection and Water Management focuses on assisting projects of national importance. The fund provides loans in two different ways: it can provide direct loans for specific projects at subsidized interest rates, or it can provide low-interest loans as 'seed money' for enterprises to provide environmental products or services. As suggested above, the National Fund is revolving; it collects revenue from interest-earnings as well as collected environmental charges.

Voivodship Environmental Funds can be provided either by loans or grants. Grants usually are restricted to monitoring activities or environmental education. Loans are provided for regional environmental investment projects, that is, environmental projects involving one or more local governments. Borrowing by local governments may be of the revenue type (where the project will realize income directly through user charges) or general obligation loans based on the taxation capacity of the local government(s). In this way, the voivodship fund is largely a revolving fund where both disbursements and earnings return as principal and interest charges. Thus, the voivodship fund has a limited earning capacity in its own right over and above its revenues from issuing permits and imposing fines.

Local Environmental Funds were introduced in 1993 as a result of heavy pressure from self-government lobbies to participate directly in the distribution of environmental funds. Under a new law passed in April 1993 (and which took effect two months later), the municipality keeps 100 per cent of the charges for removing trees and bushes (1993 Dziennik Ustaw, No. 40, item 193). Local governments also retain 50 per cent of solid waste disposal charges and 10 per cent of other environmental charges and fines for their special environmental funds. Local self-governments put the local share of such funds to use directly through local environmental investments. The funds are earmarked, and thus cannot be used for purposes other than environmental protection projects. The local governments can invest the funds directly, but more frequently they either form joint ventures with environmental firms or combine local funds with foreign assistance, voivodship funds, and loans from public and private sources.

In addition to national, voivodship and local environmental funds, two others — the Forestry Fund and the Farmland Protection Fund — are alsoregarded as ecological funds. These two funds are based on several sources of income, including charges for using forests (or farmland) for non-forestry-related (or nonagricultural) purposes. These charges may be viewed as a form of resource taxation (Żylicz, 1995). The Forestry Fund serves redistributive purposes only within the State Forest Enterprise. The Farmland Protection Fund is administered by the Ministry of Agriculture and Food Economy. It does not, however, play a significant role in resource allocation; in 1994, total revenues of the Farmland Protection Fund amounted to less than 12 million (GUS, 1995b, p. 46). Because of the limited purviews of the Forestry Fund and Farmland Protection Fund, neither is viewed as an important environmental policy instrument or as a source of funds for environmental investments.

The Bank for Environmental Protection

Another institutional innovation providing environmental funds to potential investors in Poland is the Bank for Environmental Protection (*Bank Ochrony Środowiska* or BOŚ). BOŚ was launched by a 44.4 per cent equity investment from the National Environmental Protection and Water Management Fund, with remaining capital investments coming from private Polish investors. The Bank is a full service, for-profit commercial bank which, under charter, provides preferential loans to environmental projects or to individual enterprises that provide environmental products or services. In providing preferential loans for environmental purposes, the Bank is supposed to follow national policy guidelines. Those guidelines are executed primarily by BOŚ's Board of Directors who secure the appropriate number of votes from representatives of the NEPWMF, which remains BOŚ's major shareholder. The National Fund subsidizes BOŚ's preferential loans. Those loans are expected to provide a return on equity to the National Fund, and so the BOŚ is another source of income for the Fund. The National Fund is self-sustaining thanks to its annual revenues from charges, fines, and income from its market operations, including the Bank for Environmental Protection.

Due to its modest size, the Bank for Environmental Protection cannot provide loans for all ecological projects; there is a need for other banks to enter the environmental market (Stodulski, 1994). In 1995 we interviewed the presidents of several large banks in Poland who expressed interest in providing loans for environmental projects using the same principles as the BOŚ. Unfortunately, neither the Ministry nor the NEPWMF currently seem interested in underwriting the expansion of preferential environmental loans to other commercial banks.

Debt-for-Environment Swaps and the EcoFund

One of the most innovative ventures in Poland is the debt-for-environment swap. Poland entered is transition period owing more than $32 billion to foreign governments. Poland's domestic economic crisis at the end of the 1980s made repayment of this enormous debt virtually impossible. In April 1991 the 'Paris Club', a consortium of Western creditor countries, agreed to reduce Poland's debt by 50 per cent. Between 1991 and 1993 three Paris Club member states — the United States, France and Switzerland — agreed to forgive an additional 10 per cent of Poland's debt (amounting to about $500 million) for investments in environmental projects with international significance. These funds are administered by the EcoFund, an international board headed up by Maciej Nowicki, former Minister of Environmental Protection, Natural Resources and Forestry. Poland entered into a separate debt-for-environment swap with Finland, which is managed by a special Polish-Finnish Task Force.

Grants provided by the EcoFund can cover between 10 and 30 per cent of a project's costs. This partial but significant support acts as a catalyst for mobilizing capital from other sources. The EcoFund's four main priorities reflect its obligation to finance environmental projects with international significance: (1) to reduce emissions of greenhouse gases; (2) to limit the flow of transboundary atmospheric pollution; (3) to reduce pollution of the Baltic Sea; and (4) to conserve biodiversity in Poland.

Here is one example of how the EcoFund operates. In the early 1990s the authorities decided to close the Huta Bankowa steel mill in the Katowice region for environmental reasons: the plant exceeded permitted levels of dust and gas emissions. However, the steel mill had no money to pay the fines. Despite the serious environmental problems, there were strong social reasons to keep the plant open: Huta Bankowa was a major employer, providing jobs for 2400 people in an already economically depressed area. To avoid closing, the steel mill proposed to undergo environmental and economic restructuring. The EcoFund provided a loan to replace the mill's outdated, inefficient and highly polluting boiler system. Restructuring will cost some jobs, but save many that otherwise would have been lost. Meanwhile, the EcoFund loan will lower pollution emissions from Huta Bankowa to acceptable levels.

Assessing the Self-Financing Framework

Comparing Poland today with Poland in 1990, one sees dramatic improvements. The economy has been booming since 1992, with a growth rate of nearly 7 per cent in 1995 and 1996. This is in stark contrast to many other transitional economies, particular those in the former Soviet Union, which still experiencing negative growth. Despite Poland's rapid economic growth, emissions of all major air pollutants, including particulate matter, sulphur dioxide and nitrogen oxides, have continued to decline (GUS, 1995b, p. 30). Water quality has improved significantly as pollution levels have fallen (GUS, 1995b, pp. 17-18). In addition, energy use has declined due to significant gains in energy conservation and efficiency.

All of these factors indicate that Poland is moving toward a path of sustainable development. This substantiates our earlier findings (Bolan and Bochniarz, 1994). First, there already has been progress in the transition to a market economy and a pluralistic society, which has led to economic restructuring and the dismantling of some old polluting industries. Second, new institutions for sustainable development, including the self-financing framework and strong environmental law enforcement mechanisms, have helped draft market forces to serve environmental protection.

For economists one of the most appealing indicators is the size of the environmental market in Poland. Since 1990, Poland has spent approximately one per cent of GDP per year on environmental protection, and about 0.5 per cent per year on water management (GUS, 1995B, p.45). These investments have contributed to the establishment of a $1 billion environmental market, which includes more than 1500 Polish companies and 100 foreign firms. Among the most visible signs of Poland's environmental investment campaign are wastewater treatment plants, more of which — about 1,000 — were built in the last six years than during all 45 years of communism. The same is true for the installation of air pollution control devices.

Needless to say, there is still room for further improvements. For example, collections of environmental fees and fines could be improved. In 1994, 74 per cent of assessed fees and only 13 per cent of assessed fines were collected (GUS 1995b, 46). Only 5 per cent of assessed fees for discharges of salinated water (from coal mines) were collected. On the other hand, authorities collected 90 per cent of the assessed charges for sulphur dioxide and nitrogen oxide emissions. Meanwhile, polluters that bear the brunt of environmental charges, such as the coal industry, lobby the Polish government to lower or waive fees and fines. The coal industry is one of Poland's largest employers; coal is Poland's major energy fuel and provides export revenues. However, the coal industry also is a leading loss-maker in the Polish economy. Most coal mining enterprises simply cannot afford to pay assessed environmental charges.

Despite these problems, Poland's self-financing environmental protection has certain benefits. Environmental protection tends to have more resources than other social service sectors because its funding sources are earmarked and off budget. Under the self-financing system, environmental protection is linked directly to economic conditions, and does not depend on an annual appropriation by Parliament. For this reason, environmental funding is insulated (to some extent) from political interference. This has been especially important over the past several years, during which on-budget public spending was subject to severe budget restraints. However, being off budget also has its costs. By removing substantial revenues from the budget process and the public oversight that process entails, there may be greater opportunities for mismanagement and abuse of environmental funds. This is particularly a problem for the voivodship and local environmental funds which are subject to less oversight than the National Fund or the Environmental Protection

Bank.[5] Similar quasi-public funds in other countries have been subject to abuse.

Finally, Poland's new institutional system is premised on some key assumptions about the future. One is that the ratio of nonperforming investments to total investments will be tolerably low; bad environmental loans should be expected to have the same effects as bad loans generally have in market economies. So far, however, environmental loans in Poland have tended to perform better than general (non-environmental) loans from Polish commercial banks.

Summary

Poland has created a kind of quasi-market focusing on environmental protection and clean-up. The environmental fee has become an instrument for capital formation. This capital is then available, through something like a competitive bidding process, for environmental investments. The National Fund, the voivodship funds, and the Bank for Environmental Protection are the trustees and allocators of this environmental capital. Projects are selected for investment based on their potential for both financial and social (in the form of environmental protection and restoration) return. This is a quasi-market in the sense that it depends on subsidized rates of interest,[6] and its social goals are at least as prominent as its financial goals.

The Polish system of self-financing environmental protection provides a model for the development of economic instruments for environmental protection in other countries. But Poland's experiences also demonstrate that, under current economic conditions, such a system alone does not (yet) provide a sufficient institutional basis for sustainable development. Only when every nation provides true price signals for water, air, minerals, flora, and fauna can any one nation fully build toward sustainable development. Until that point is reached, however, Poland provides a useful example of how an imperfect quasi-market can be effective in generating both incentives and resources for environmental protection and restoration.

Notes

1. The environmental economist Tomasz Żylicz (1994) has noted three basic ideas in the Polish policy. The first is the 'polluter pays principle'. The second is decentralization (sub-national regionalism) that makes local authorities responsible for compliance with national standards within their jurisdictions. And the third is efficiency, which emphasizes the need for the flexibility of the market system in allocating abatement activities. Żylicz also suggests that to achieve an assumed goal of reducing sulphur dioxides by 30 per cent, current fees would have to be increased sixfold to equal abatement costs.
2. Actually, Poland's extensive use of environmental charges substantially predates the transition era. In People's Poland, resource use and pollution fees were imposed beginning in the 1960s, and by the 1980s were a primary mechanism of environmental protection. However, the levels of environmental charges in communist Poland were far too low to deter noncompliance. The dramatic fee increases of the early 1990s therefore reflect a new commitment to effluent and resource use charges as a basic mechanism of environmental protection.
3. There is also a State Epidemiological and Sanitary Inspectorate which monitors pollution from the perspective of its human health impacts.
4. Fees for mineral extraction are split evenly between the municipality where the extraction occurs and the National Fund. Significantly, local governments may treat these fees as general revenues useable for any purpose. All other environmental revenues, by contrast, are earmarked for environmental investments.
5. Under the April 1993 legislation (discussed earlier) establishing the local environmental protection funds, the voivodship funds were restructured along the lines of the National Fund. Each voivodship fund now has an independent Board appointed to oversee fund operations. This presumably reduces the opportunity for mismanagement and corruption by local officials.
6. The subsidies might spur other capital markets in the following way: If the environmental funds grow large enough, their subsidized interest rates might increase the competition for capital, thereby driving down interest rates and, in turn, increasing the availability of venture capital. However, this is only likely to occur if strategic behavior (most notably free-riding) is eliminated from the environmental capital funds.

References

Bolan, Richard and Zbigniew Bochniarz (1994), *Poland's Path to Sustainable Development 1989-1993*, Hubert H. Humphrey Institute for Public Affairs, Minneapolis.

Cole, Daniel (1995), 'Poland's Progress: Environmental Protection in a Period of Transition', *The Parker School Journal of East European Law*, vol. 2, pp. 279-319.

GUS (1995a), *Ochrona Środowiska 1995*, Głowny Urząd Statystyczny, Warsaw.

GUS (1995b), *Rocznik Statystyczny 1995*, Głowny Urząd Statystyczny, Warsaw.

Stodulski, Wojciech (1994), *The case of banking and the environment in Poland, Report no. 3*, Institute for Sustainable Development, Warsaw.

Żylicz, Tomasz (1995), 'Pollution Taxes as a Source of Budgetary Revenues in Economies in Transition', in L. Bovenberg & S. Cnossen (eds), *Public Economies and the Environment in an Imperfect World*, Kluwer, Boston, pp. 203-218.

Żylicz, Tomasz (1994), 'Environmental Policy Reform in Poland', in T. Sterner (ed), *Economic Policies for Sustainable Development*, Kluwer, Dordrecht, pp. 82-112.

3 Economic Instruments in Polish Environmental Policy

JERZY ŚLESZYŃSKI

Introduction

Poland, like many other countries, has for years relied on a mixed system of standards and fees to address environmental problems. Poland's water resources management system has relied on individual environmental fees since the 1960s. In the 1970s environmental charges became an integrated part of the national environmental management system. However, within socialism's centrally planned economy, the fees were merely a façade, introduced only to demonstrate the government's commitment to environmental protection. By design, however, pollution and resource use fees could not motivate polluters to adjust their behavior to comply with environmental protection requirements. Industrial and municipal economic agents operated under so-called soft budget constraint, a common feature of Soviet-style economies, which undermined the effectiveness of fees and fines up until the time Poland began the transition to a market economy. Financial instruments for environmental protection were doomed to fail in an economy in which all essential inputs were allocated administratively and plant managers had little incentive to pay attention to price stimuli. Moreover, initially, Poland's environmental fees were set rather low; even if polluters and resource users had been sensitive to price stimuli, the fees would have had little effect.

During the 1970s and 1980s, however, the Polish government revaluated its environmental fee schedules several times; the fees reached their peak pre-reform level in 1987. But in the absence of real market conditions, even relatively high fee levels made little difference. As Poland began its transition to market democracy in 1989, its environmental policy changed dramatically (National Environmental Policy, 1990). Since 1990 Poland has implemented a series of significant environmental fee increases. Today the fees are quite high, and generate substantial revenues that are recirculated into the economy to finance environmental improvements. The January 1995 fee increases were intended to avoid the devaluation of the fees by Poland's high inflation rate. Overall, environmental fees have risen to approximately 18-20 times their levels during the Communist regime.

Pollution and resource use fees play key roles in Poland's system of environmental protection. They apply to a variety of economic activities that create air pollution, use water, discharge waste water, dispose of solid waste, cut trees and bushes, and generate noise pollution. Poland's current pollution fees are comparatively high; several of them rank among the highest in the world (e.g., Poland's SO_2 emission fee as of January 1996 was US$/ton 94). Economic instruments other than pollution and resource fees are of little importance under present regulations. However, there is growing interest in product fees and tradeable permit policies, which currently are being researched and officially considered by the government.

This paper focuses on the implementation, performance, and enforcement of pollution fees. The analysis is supplemented with information concerning non-compliance fines and subsidies. Certain other non-environmental economic instruments with discrete 'green' provisions are discussed in passing.

Environmental and Resource Fees

Implementation

Environmental fees and fines in Poland's environmental policy are closely linked to and complemented by the emission and discharge standards that are included in each individual plant's operating permit. The permits specify allowable discharges of every regulated pollutant from every pipe, stack or other type of conveyance. Local environmental authorities are supposed to set allowable individual emission and discharge levels so as to achieve or maintain ambient air and water quality standards.

Poland's pollution and resource use fees operate in a mixed system in which the principle of command-and-control plays the leading role. The legal and administrative regulations (command-and-control), in the form of emissions, pollution and resource use fees, which provide a lower-cost means of addressing specific environmental problems.

In general, both the *User Pays Principle* and the *Polluter Pays Principle* provide a rationale for the fees. Today fee levels are determined on the basis of self-reported emission and discharge amounts. This marks a great advance over the old system, in which fees were based only on the emission or discharge levels specified in operating permits.

Relative differences in fee rates correlate with differences in the toxicity or potential to cause environmental damage of different pollutants. The absolute magnitudes of the rates, however, do not reflect marginal damages to health and the environment or correspond with the marginal costs of abatement. There are two other reasons why fee rates differ:

- water resources conditions vary by geographical region — fees for surface and ground water extraction are higher in the regions in which water is in short supply;
- uses vary — fees charged to 'dirty' industries for surface and ground water extraction and waste water discharges are higher and, in general, industry pays much more than the public sector.

It is particularly important to bear in mind that enterprises can treat environmental fees as production costs because they are deductible from taxable income.

Performance

Environmental fees are collected once a year by the provincial (Voivodship) environmental protection departments. For reasons of administrative efficiency, large enterprises pay their fees in quarterly installments.

The use of revenues from fees and fines is legally restricted to ecological purposes. In other words, they are off-budget. This distinguishes Poland from other countries-in-transition where collected environmental charges become general budget revenues available for non-ecological investments. Because Poland's pollution and resource use fees and fines are earmarked for environmental investments, those investments are protected against efforts to cut programs and spending to generate budget 'savings'.

Revenues from pollution fees, resource use fees, and non-compliance fines are distributed to the National Fund for Environmental Protection and Water Management, 49 provincial (Voivodship) environmental funds and 2,468 municipal (*Gmina*) environmental funds. The pattern of distribution among this group of recipients varies with respect to the nature of the activity being charged, as designated in regulations. Sometimes, the distribution is based on contradictory factors. But it is also related, to some extent, to political bargaining over the decentralization of environmental responsibilities and decisionmaking. In general, approximately 10 per cent of all collected environmental fees and fines remain in municipal environmental funds, while 54 per cent goes to the provincial environmental funds and 36 per cent to the National Fund. This pattern applies to air pollution fees, waste water fees, and fees for water consumption. A more detailed description of major financial flows is presented in Table 3.1.

Table 3.1 Major Distribution Schemes for Collected Environmental Fees and Fines

Collected fees and non-compliance fines on SO_2 air pollution; Fees on surface and ground water withdrawals; Fees and non-compliance fines on waste water discharges:

36% goes to the National Fund
54% goes to Provincial Environmental Funds
10% goes to Municipal Environmental Funds

Collected fees and non-compliance fines on NO_x air pollution; Fees and non-compliance fines on saline water discharges:

90% goes to the National Environmental Fund
10% goes to Municipal Environmental Funds

Collected fees and non-compliance fines on waste disposal:

20% goes to the National Environmental Fund
30% goes to Provincial Environmental Funds
50% goes to Municipal Environmental Funds

Collected fees on cutting trees and bushes:

100% goes to Municipal Environmental Funds

Collected minerals extraction fees; Fees on mining concessions:

40% goes to the National Environmental Fund
60% goes to Municipal Environmental Funds

Tax differentiation on gasoline:

100% goes to the State Budget

Source: Spyrka, 1994; supplemented by the author.

Revenues from sulphur dioxide emissions, nitrogen oxide emissions, and mining waste water discharges are earmarked for investments and other expenditures intended to reduce those specific pollutants. However, it is extremely difficult to address specific types of pollution separately. Thus in practice selected investment projects tend to address multiple environmental protection issues, but with an emphasis on sulphur dioxide, nitrogen oxides or waste water from mining.

Since 1992 fee rates have been revised annually to maintain the real rate structure; nominal fee rates are adjusted each year to account for inflation. At first, because the rate adjustments accounted only for the preceding year's inflation, the fee rates could not keep pace with inflation. This weakened the effect on polluters, reducing their incentive to change their production patterns. More recent fee adjustments have secured the real level of fee rates and revenues by accounting for projected inflation through the coming year.

Poland's fee rates are high when compared to OECD standards, but are not high enough to stimulate polluters to undertake environmental investments sufficient to meet emission or discharge targets. Aside from a few cases, such as old power plants, in which environmental expenditures make up a large part of total operating costs, it is generally the case that rates fall short of the so-called Pigouvian level. For instance, the estimated marginal cost of investment to achieve a 30 per cent reduction in SO_2 would require US$ 600 per ton for large combustion sources (Anderson and Fiedor 1995). This value is seven times higher than Poland's current fee level for SO_2 emissions, which is the highest in the world. Given Poland's present conditions, further significant increases in fee rates are unlikely and would be questionable as a matter of policy.

Fee and fine collection rates have continually declined as the total of assessed fees and fines has increased (Table 3.2). However, this trend may have begun to improve in 1994. Collection efficiency is illustrated in detail (with respect to different fees from 1990-1994) in the tables on fee rates and raised revenues.

Moreover, the statistics on collection efficiency in 1994 clearly indicate that air pollution and water withdrawal fee collections are very efficient (Table 3.3). Collection efficiency is less satisfactory for waste disposal, and only 50/50 for waste water fees and tree cutting fees. It is critically low for non-compliance fines, and especially critical for saline water fees collected from coal mining sector.

Table 3.2 Revenues Provided by Environmental Fees and Fines (US$ 1,000,000)

	1990	*1991*	*1992*	*1993*
Imposed fees and fines	31.1	523.1	649.6	660.6
Amounts actually collected	30.1	385.7	446.8	428.5
Collection efficiency rate	96.8%	73.7%	68.8%	64.9%

Source: Kruszewski, 1994.

Table 3.3 Revenue from Environmental Fees and Fines in 1994

Type of instrument	*Amounts imposed (Mill. US$)*	*Amounts collected (Mill. US$)*	*Collection efficiency rate*
Environmental fees (total)	585.9	431.6	74%
of which:			
- water withdrawal	56.9	54.6	96%
- waste water	200.9	91.3	45%
discharge	245.8	222.0	90%
- air pollution	81.9	63.5	77%
- waste disposal	0.2	0.1	47%
- tree cutting			
Non-compliance fines	101.4	13.5	13%
Minerals extraction fee	n.a.	41.2	n.a.

Source: Informacja..., 1994; Anderson and Fiedor, 1994.

Enforcement

In addition to direct regulation and its enforcement mechanisms, economic enforcement tools play an important role. Non-compliance fines are penalties levied on the quantity of emissions or discharges in excess of the allowable limit specified in each facility's permit. In contrast to pollution and resource use fees, enterprises are not allowed to treat non-compliance penalties as deductible production expenses.

As indicated in the previous section, the collection rate for non-compliance fines is very poor. There are two reasons for this. First, a number of enterprises in chronic non-compliance will be closed very soon for economic reasons; they have no prospects for modernization or privatization, or for sorting out their environmental liabilities. Second, several other heavily polluting enterprises are now in the process of overcoming their economic troubles through modernization or restructuring. During this process, the payment of fines, including environmental non-compliance fines, is deferred. Deferred fines may be reduced or forgiven entirely if enterprises undertake environmental investments to permanently reduce excessive pollution. While this policy contributes to the low collection rate of environmental fines, it may benefit environmental protection overall.

Much more troublesome is the limited ability of regional environmental protection inspectors to monitor and verify emission and discharge amounts: they have too few resources and there are too many facilities to monitor and inspect. In addition, some analysts estimate that nearly half of all polluting facilities in Poland operate without valid permits, since the limited resources of local permitting authorities has created a backlog in the permitting process. However, the extent of this problem should not be exaggerated. Most firms operating without permits are small and minor polluters. Moreover, a more realistic analysis of pollution sources operating without permits would likely show that estimates of the number of sources requiring permits should be revised downward.

Approximately 800 major polluters in Poland are subject to special management supervision by the provincial offices of the State Environmental Protection Inspectorate. At present, 74 enterprises remain on the official list of the worst polluters which is officially published for public inspection and comment. According to a recent OECD overview of environmental law enforcement in Poland, approximately 70 per cent of former violators are now in compliance with their permit requirements (Anderson and Fiedor, 1995).

Some technical obstacles to enforcement result from limited financial resources and years of neglect. The national environmental monitoring network for air, noise, surface and ground waters, nature protection, and radioactive contamination is expected to be completed by 1998. Until then, only partial and incomplete monitoring will be available.

Fee Levels

The Polish system of environmental fees is quite well established relative to fee systems in other Central and Eastern European countries. In principle, the system complies with OECD guidelines. The Polish system's level and scope of fees may be considered to be an advanced approach to pricing the environment. On the other hand, the Polish system's performance suffers from various problems that need to be analyzed and eliminated. Compared to other countries, Poland's environmental fees are widely accepted; they enjoy significant public support. In general, at Poland's present stage of economic transformation, the fee system can be said to perform satisfactorily.

Environmental fees have set rates for water and air pollution, solid waste disposal, and for water consumption. In addition, natural resources users are charged for tree cutting, for water transport, for mining concessions, for converting forest land to non-forest uses, and for converting agricultural land to non-agricultural uses.

In the case of air pollution, emission fees are levied on emissions of 62 atmospheric pollutants according to a complicated formula. In 1996 the rate structure distinguished three categories of pollutants with strongly differentiated fee levels, ranging from 0.13 to 158.08 new Polish zlotys[1] per kg of pollutant. For example, the fee rates for hydrocarbons and particulate matter are significantly differentiated among each category. The highest rates apply for toxic emissions like benzene and chromium. In particular, emission fees for the most important gaseous pollutants, SO_2 and NO_x, are among the highest in the world at 0.24 new zlotys per kg.

Revenues from air pollution fees are quite high thanks to exceptionally strong collection rates. For instance, revenues from sulfur dioxide emissions fees have increased significantly year-by-year — in 1996 they were almost 50 per cent higher than in 1991; collection efficiency of 97 per cent provided revenue in 1994 equal to US$ 105.24 million.

Fees on nitrogen oxide emissions are exactly the same as for sulfur dioxide. However, the revenues raised are significantly lower and collection efficiency slightly worse.

Charges for discharging waste water and saline water are based on BOD5, COD, suspended solids, heavy metals, and chlorate and sulphate ions content. Chlorate and sulphate ions in saline water are charged differently, depending on whether they are discharged directly to the surface water or discharged to dosing aquifers. Besides the high rates charged for BOD5 and COD, users are charged for volatile phenols discharges. BOD5 and COD rates are differentiated by economic sectors. The BOD5 rate ranges from 0.51 new zlotys per kg for rural municipal sewage, hospitals and social care institutions, to 5.06 new zlotys per kg for power generation, fuel processing, chemical, metallurgical, machine and light industries. The COD rate varies, respectively, from 0.28 to 3.55 new zlotys per kg. The waste water and air pollution fees are the same for all administrative regions.

Environmental fees are also levied on surface water and groundwater withdrawals. The base rate for surface water is 77.00 new zlotys per 1000 m^3, and for groundwater 243.00 new zlotys per 1000m^3. These rates are differentiated by economic sector and administrative region. Furthermore, water users pay user fees to cover investment and maintenance costs of water supply installations, as well as sewage collection and treatment. However, these costs are partially subsidized by budget allocations in some municipalities.

The waste disposal fee distinguishes among 163 substances in four categories. The waste categories are determined by toxicity, and separate rates apply for each category. The fees range from 4.75 to 63.23 new zlotys. There is also a municipally administered fee for tree removal which is differentiated according to the species and size of the tree.

In addition to the fees already discussed, there are fees for mineral extraction and mining concessions. The extraction fees are based on the prices of the minerals. In 1992 the National Fund of Environmental Protection started receiving a portion of mining fees, most of which are recirculated to the mining sector to finance exploration and prospecting as well as environmental protection. The mining fees play an important revenue-raising role: given the substantial role of the mining sector in Poland's economy, even modest and inefficiently collected fees provide as much as 12 per cent of National Fund revenues. This significant sum should help speed up the implementation of environmental projects at coal mines. These fees are too low, however, to provide a significant incentive for conserving exhaustible resources.

Other Economic Instruments

There are not, in fact, many different kinds of economic instrument in Polish environmental policy. The dominant role of environmental and resource fees is obvious. Other instruments play only marginal roles and do not significantly influence the level of environmental expenditures either of business or the public sector.

A differentiated tax on gasoline was designed to promote unleaded gasoline sales. Unfortunately, the difference reflected in price is not significant at all, so it has no effect on present patterns of gasoline consumption.

A deposit-refund system for beverage containers is still on the initial stage of its development in Poland. On the one hand, deposits on beverage containers are not high enough to attract consumers for effective collection. On the other hand, the system covers only a small range of all glass and plastic containers being produced, imported, and deposited in Poland.

Table 3.4 Emission Fees for Atmospheric Pollutants

	Pollutants emitted to the atmosphere	*Rates (US$ per ton)*
Group I (selection):		
-	acrylonitrile (aerosol), arsenium*, asbestos, benzene, benzo(a)pyrene, chlorinated vinyl (gaseous), chromium*, nickel*	61,968
-	bismuth*, cerium*, CFCs (CFC-11, CFC-12, CFC-13, CFC-111, CFC-112, CFC-113, CFC-114, CFC-115, CFC-211, CFC-212, CFC-213, CFC-214, CFC-215, CFC-216, CFC-217), dioxines, halons (1211, 1301, 2402), cadmium*, cobalt*, manganese*, molybdenum*, lead*, polychlorinated biphenyls, mercury*	30,984
-	tin*, zinc*	3,098
Group II (selection):		
-	particulate matters (except those specified somewhere else), hydrocarbons (except those specified somewhere else)	247
-	sulphur dioxide, nitrogen oxides (calculated as NO_2)	94
-	polymers particulate matters	86
-	particulate matter emitted from combustion processes	51
-	carbon monoxide, aliphatic hydrocarbons and their derivatives	24
-	carbon dioxide, methane	0.051

Group III (selection):

-	inorganic acids and their derivatives (except those specified somewhere else)	337
-	aliphatic aldehydes and their derivatives	165
-	HCFCs, halons (except those specified somewhere else), oils fog	86

* fee rate applies to the metal content

Table 3.5 Fees for Waste Water (rates in US$ per ton)

Pollution sources	*BOD_5*	*COD*	*SS*	*CSI*	*HM*	*VP*
Power generation, fuel processing, chemical, metallurgical, machine, and light industries	1,984	1,392	85	7	9,918	3,720
Paper and pulp industries	843	501	85	7	9,918	3,720
Food industry	496	334	85	7	9,918	3,720
Municipal sewage systems, sewage from health and social care institutions	198	111	85	7	9,918	3,720
Other sources (except saline coal-mining waters)	992	557	85	7	9,918	3,720
Saline coal-mining waters discharged directly to an aquifer	-	-	-	55	-	-
Saline coal-mining waters discharged from dosing reservoirs	-	-	-	7	-	-

BOD_5 biochemical oxygen demand during the first 5 days
COD chemical oxygen demand
SS suspended solids
CSI chloride and sulphate ions
HM heavy metals (total mass)
VP volatile phenols

Table 3.6 Fees for Surface Water Withdrawal (US$/1000 m³)

Users of the surface inland water resources	*Katowice Province*	*Provinces: Bielsko-Biała, Częstochowa, Jelenia Góra, Kielce, Kraków, Legnica, N. Sącz, Opole, Tarnów, Wałbrzych, Wrocław*	*Other 37 provinces*
Industrial uses connected with electricity and heat production	30.18	15.09	6.04
Other industrial uses	150.92	75.46	30.18
Household, municipal, and selected public sector uses, agriculture and cultivation	5.10	5.10	5.10
Irrigation in agriculture and forestry	4.00	4.00	4.00

Table 3.7 Fees for Ground Water Withdrawal (US$/1000 m³)

Users of the ground water resources	*Provinces: Bielsko-Biała, Chełm, Częstochowa, Jelenia Góra, Katowice, Kielce, Kraków, Legnica, Lublin, Nowy Sącz, Opole, Tarnów, Wałbrzych, Wrocław*	*Other 35 provinces*
Industrial uses connected with food and drugs production	76.21	28.58
Other industrial uses	238.14	95.26
Household, municipal, and selected public sector uses, agriculture and cultivation	5.10	5.10
Irrigation in agriculture and forestry	4.00	4.00

Table 3.8 Fees for Waste Disposal

Categories of disposed waste	*Rates (US$ per ton)*
Group I: 62 most toxic wastes, including e.g. waste with high heavy metal concentration, used lubricants, asbestos, hospital waste, sludge from primary treatment of urban municipal sewage	24.79
Group II: 45 medium toxic wastes, including e.g. waste with low heavy metal concentration, polychlorinated biphenyls, used pharmaceuticals, rotten foodstuffs and feedstock	9.29
Group III: 36 low toxic wastes, including e.g. ashes, flue gas desulphurization sludge, fruit and vegetable processing waste	3.10
Group IV: 20 non-toxic wastes, including e.g. coal-washing sludge, glass, paper, building demolition debris	1.86

Table 3.9 Tax Differentiation on Gasoline (Level: May, 1996)

Leaded gasoline:
0.21 US$/liter (ca. 38% of the retail price)

Unleaded gasoline:
0.19 US$/liter (ca. 35% of the retail price)

Source: Spyrka, 1996.

Anticipating that fee decentralization was inevitable, the Ministry of Environmental Protection in May 1990 tried to introduce a fuel fee on top of existing excise taxes. The proposed fee would have averaged 4 per cent, with revenues earmarked for environmental protection. However, the proposal was defeated by almost unanimous opposition from political groups. Another attempt to enact a similar tax a few years later met with equally strong criticism. Consequently, Poland's environmental policy presently includes no product fees, which is especially curious considering Poland's progress in funding environmental protection and European trends in product fee promotion.

Table 3.10 Deposit-Refund on Beverage Containers (Deposit rates: May, 1996)

0.50 l beer bottles (glass):
0.15 US$/bottle (14%-20% of the product's market price)

0.25 l soft drink bottles (glass):
0.07 US$/bottle (21%-27% of the product's market price)

0.33 l soft drink bottles (glass):
0.07 US$/bottle (20%-24% of the product's market price)

1.00 l soft drink bottles (plastic):
0.19 US$/bottle (21% of the product's market price)

1.50 l soft drink bottles (plastic):
0.22 US$/bottle (19% of the product's market price)

2.00 l soft drink bottles (plastic):
0.37 US$/bottle (23%-34% of the product's market price)

Source: Spyrka, 1996.

Other Taxes and Duties

Additional fiscal measures are employed to encourage environmental investments, recycling, and consumption of 'green' products. Generous tax allowances for environmental investments were discontinued in 1991. Nevertheless, there are still some instruments which marginally improve the profitability of recycling, conservation, and other environmental measures (Żylicz, 1996):

- the value added tax rate is lower for products related to environmental protection;[2]
- donations for environmental purposes are deductible from income tax;
- donations of real estate received for environmental protection purposes are exempted from income tax;
- income tax allowances are granted to farmers who invest in or

modernize environmental protection equipment, renewable energy sources, installations to enhance the use of water;

- exemptions from real estate tax are given to water supply and sewage collection installations, land used for water retention reservoirs and hydro-power stations, various categories of forest land;
- sales regulated by the Environmental Protection Act are exempt from real estate sales tax;
- non-profit organizations active in environmental protection are exempt from administrative fees and income tax; and
- farmers receive an interest rate subsidy for adapting their farms to ecological farming principles.

Only one excise tax in Poland has an environmental objective. A plastic container tax, introduced in July 1995 by Decree of the Minister of Finance, subjects packaging produced from polyethylene and polypropylene to 10 per cent of the regular tax rate. Packaging produced from polycarbons, polystyrene, and PET is subject to 20 per cent of the regular rate. Regular rates for the 13 types of containers range from 1 new zloty per 1000 pieces for small plates, to 0.03 new zloty per 1.5 liter bottle, and to 6 zlotys for a drum.

Liquid fuels are subject to fairly high excise taxes. Taxes on petrol are now in the order of 40 per cent of retail prices.

Tariffs on automobile imports rise with car age. This should encourage the import of newer, less polluting cars. There is a complete ban on the import of cars over 10 years old.

A number of other fiscal measures act as environmental taxes or subsidies. Although they do not create much of an incentive to reduce pollution or increase conservation, they do generate revenues for environmental investments and activities. Since these measures can be seen as indirect subsidies, their efficiency demands serious analysis. Because many of them were introduced either on an ad hoc basis, as green ornaments to various bills and regulations or as quick compromises between various governmental departments, their equity and efficacy are open to question.

Subsidies

The most distinctive feature of environmental management in Poland is the extensive system of pollution and natural resource fees collected and redistributed for abatement and conservation purposes through special funds: the National Fund for Environmental Protection and Water Resource Management (National Fund), 49 provincial environmental protection funds, and, since 1993, more than 2,400 municipal environmental funds. These funds primarily subsidize environmental investments through grants and soft loans.

In general, 10 per cent of all collected environmental fees and fines remains for the municipal funds, while 54 per cent goes to the provincial funds and 36 per cent to the National Fund. A more detailed description of the financial flows was presented in Table 3.1.

The recent growth in incremental environmental expenditures can be attributed to the rapid increase in pollution fees and non-compliance fines which are recirculated through the National Fund, and its provincial and municipal counterparts. The estimated breakdown of financing the overall environmental investment is presented below. Similar patterns of expenditures were estimated to prevail in the whole three-years period (Table 3.13). It should be noted that 95 per cent or more of all environmental expenditures in Poland have been financed from domestic sources. However, the state budget's contribution to those expenditures has been quite marginal.

In 1992, despite problems with maintaining the initially planned level of fees, environmental funds managed to collect and spend almost 40 per cent more than in 1991 (Anderson and Żylicz, 1993). The high level of investment has been sustained in real terms, though there have been no further significant increases. The environmental funds' share in environmental investment in 1992 was 58 per cent. On the one hand, this underscores the critical role of Poland's pollution fee system in the environmental recovery process. On the other hand, however, it reveals the fragility of environmental protection enforcement. Namely, the direct expenditures from polluters own sources decreased immediately in response to the political turbulence resulting in the decline of competence and prestige of the Ministry of Environmental Protection in 1992. In 1993 the contribution of all environmental funds to environmental investments declined to 47 per cent, and to 40 per cent the following year.

Expenditures on water protection have begun to rise after an initial decline, while air pollution expenditures have stabilized over the last three years after early growth (Table 3.14). In 1993, 52 per cent of all environmental expenditures went to water protection and management, 39 per cent to air protection, and only 9 per cent to solid waste management.

Total expenditures on environmental protection in 1991 were almost 45 per cent higher in real terms than in 1990 (the highest previous level). As a percentage of the country's national income, environmental expenditures grew from 0.7 per cent of GDP in 1990 to 1.1 per cent of GDP in 1991. That year, for the first time in terms of percentage of GDP, environmental investment levels in Poland were commensurate with levels in OECD countries. In succeeding years, the level of environmental investments grew even higher.

The 1993 environmental spending level was almost double the 1989 level.

Table 3.11 Revenue Sources for Environmental Investment (%)

Investment Sources	*1991*	*1992*	*1993*	*1994*
State Budget (grants)	5%	5%	7%	5%
National Environmental Fund (grants, soft loans)	15%	27%	21%	22%
Provincial Environmental Funds (grants, commercial loans)	25%	31%	25%	15%
Municipal Environmental Funds (grants)	-	-	1%	4%
Enterprises (own resources, commercial loans)	30%	20%	25%	31%
Municipalities (budgets, commercial loans)	20%	13%	16%	19%
Foreign Assistance (grants, development loans)	5%	4%	5%	4%

Sources: Informacje..., 1994; Ochrona..., 1995.

Table 3.12 Environmental Expenditures by Sector (%)

Sector	*1988*	*1989*	*1990*	*1991*	*1992*	*1993*	*1994*
Water	64	62	48	45	47	52	47
Air	18	18	31	40	36	39	44
Solid waste	19	20	21	15	17	9	9

The total expenditures include also expenditures on nature conservation, anti-noise measures, etc., which account for less than 1 per cent. Source: Żylicz and Lehoczki, 1994; Informacje..., 1994.

In general, the share of environmental investments in GDP (in 1990 prices) remains below the desired level of 1.5 per cent. However, it has been growing steadily. In 1992 and 1993 it was up to 1.3 per cent, and in 1994 it rose to 1.4 per cent of GDP (calculated in 1990 prices). It is interesting to note that the same indicator provides a much less optimistic interpretation when calculated in current prices.

COWIconsult has conducted research on total environmental expenditures including investment, research and development, and current costs (including administration costs). The total expenditure level for 1993 was estimated to be between 3.8 and 4.4 billion US dollars (between 2.1 and 2.5 million new Polish zlotys). Expenditures of the public sector and business sector are separated in Table 3.17. A comparison of two approaches — abatement principle and financing principle — shows that the business sector pays more in fees than it receives back in subsidies. The opposite is true for the public sector, which abates more than it pays. Therefore, it seems justified to conclude that the public sector receives a net subsidy derived from business expenditures.

Table 3.13 Total Environmental Investment

Environmental Investment	*1990*	*1991*	*1992*	*1993*	*1994*
Totals in million PLN (current prices)	415	840	1197	1512	2128
Totals in million US$ (current prices)	437	794	878	833	937
Totals in million PLN (1990 prices)	415	592	710	714	760
As a per centage of GDP (calculated in 1990 prices)	0.7	1.1	1.3	1.3	1.4
As a per centage of GDP (calculated in current prices)	0.7	1.0	1.0	1.0	1.0

Sources: Ochrona środowiska, 1995; Central Statistical Office estimates and own calculations.

Table 3.14 Environmental Investment Indicators (calculations based on data in current prices)

Indicators	*1989*	*1990*	*1991*	*1992*	*1993*	*1994*
Environmental investment/Total investment (%)	2.9	3.6	5.0	5.9	6.1	6.4
Environmental investment/capita (PLN)	1.4	10.9	21.9	31.2	39.3	54.0
Environmental investment/capita (US$)	10	11	21	23	22	24

Source: Case Study..., 1995.

Table 3.15 Environmental Expenditures in 1993 by Sectors (millions of new Polish zlotys PLN at current prices)

Expenditures	*Public Sector*	*Business Sector*
Investment	548.9	963.1
Current expenditure	268.4 - 368.4	300.0 - 550.0
Research and Development	33.2	n.a.
Abater Principle Expenditures	850.5 - 950.5	1,263.1 - 1,513.1
- subsidies to business	(+)547.0	(-)547.0
- fees/fines from business	(-)710.7	(+)710.7
Financing Principle Expenditures	686.8 - 786.8	1,426.8 - 1,676.8

Source: Case Study..., 1995.

A separate and distinct public entity, the EcoFund, plays a special role in Poland's system of financing environmental protection. It does not yet provide much financial assistance, but the concept itself opens new doors for funding opportunities, while helping to alleviate Poland's external debt.

In the years 1971-1978, Poland contracted debts worth US$ 18 billion to 17 countries. The inefficiency of the centrally planned economy meant that the debts were not repaid, and in 1991 they amounted to US$ 32 billion, including interest. Poland's political and economic transformation led the Paris Club of sovereign creditor countries in 1990 to reduce the debt by 50 per cent, provided that Poland discharge the remaining part in annual allotments by the year 2010. This agreement between Poland and the Paris Club also raised the possibility of supplementary debt-for-nature swaps between individual Paris Club member states and Poland. If sovereign creditors would forgive an additional 10 per cent of Poland's debt, the Polish government would invest a comparable sum in environmental protection projects with international significance. In 1992 Poland's Minister of Finance established the EcoFund as an independent non-profit foundation to manage and distribute resources pursuant to debt-for-nature swap arrangements with Paris Club creditors. So far, Poland has signed debt-for-nature swap agreements with the following sovereign creditors:

- United States - 10 per cent of debt, US$ 370 million;
- Switzerland - 10 per cent of debt, US$ 66 million;
- France - 1 per cent of debt, US$ 63 million;
- Finland - 10 per cent of debt, US$ 14 million.

According to the present agreements, the EcoFund will handle a total amount of US$ 500 million by the year 2010. The money will be transferred from the Polish state budget in quarterly and/or semi-annual allotments. In 1994 the income of the EcoFund amounted to US$ 8.5 million (PLN 19.4 million). The resources from the debt-for-nature swap are channeled to projects in one of four areas of international importance:

- Reduction of transboundary transport of sulphur and nitrogen oxides from the Polish territory;
- Reduction of Baltic Sea pollution;
- Reduction of greenhouse gases emission and pollutants causing the depletion of the stratospheric ozone layer;
- Protection of biological diversity.

In 1994 total grants approved amounted to $14.3 million. The EcoFund spent $3.5 million on transboundary air pollution control, $5.2 million on cleaning up the Baltic Sea, and $3.8 million on climate change programs. Around $1.7 million were spent on projects that were exclusively for nature conservation.

Non-compliance Fines

Poland has a system of non-compliance fines for all discharges above legally or administratively set limits. In contrast to the regular fees, these non-compliance fines are payable from after-tax income. Therefore, these instruments have a stronger incentive effect. Non-compliance fines are assessed on all the same pollutants for which regular fees are collected. There are also non-compliance fines for tree removal and for exceeding allowable noise levels.

The non-compliance fine for air pollutants, charged per unit above maximum allowable emissions, is 10 times the regular fee. Polluters who emit without a valid permit are charged double rates for each unit of emissions.

In the case of water pollutants, a very complicated formula is applied which results in fines on the order of the regular fees. In addition, violations of allowable temperature of discharged cooling waters, allowable acidity/alkalinity of effluent, and allowable level of radioactivity of a water permit result in non-compliance fines. Significantly, municipalities pay much lower fines than private dischargers.

Non-compliance fines for waste disposal are related to the regular fees. They are assessed as 5 per cent of the latter per every day of depositing the waste without a valid permit.

The non-compliance fines applied in the heavily-polluted administrative region of Katowice, like the regular fees, used to be twice as high as those elsewhere. In the nearly-as-polluted Kraków administrative region, only air pollution non-compliance fines were twice as high. These surcharges were discontinued in mid-1992, after the Constitutional Tribunal ruled that they were unconstitutional.

In cases involving chronic non-compliance with air and noise pollution permits, where ‘chronic’ is defined as lasting for more than three years, the fines are doubled.

Penalties can be deferred for three to five years (Anderson and Fiedor, 1995). To qualify for this deferral, enterprises must implement measures to attain compliance status. If the enterprise is successful in attaining compliance, all or a portion of the deferred penalties may be forgiven, depending on the level of expenditures the enterprise makes to attain compliance. If these investments exceed the amount of accumulated penalties, the entire penalty is canceled. If, however, the enterprise fails to attain compliance, the penalties are reinstated and a 50 per cent surcharge is added. The Environmental Protection Inspectorate can require the immediate payment of deferred penalty if the enterprise fails to take actions that would lead to compliance.

Strengths and Weaknesses of Poland's System of Economic Instruments for Environmental Protection

A system of environmental fees was first established under Poland's previous system of central economic planning. But under that system, the fees had little impact on environmental quality. More recently, the combination of fee increases and market reforms have substantially improved the effectiveness of this economic instrument. Compared with fee levels in the OECD countries, Poland's environmental fees are high and its fee structure is complex. There are literally hundreds of rates which vary across pollutants and polluter types. Although in principle differentiations are based on environmental and health considerations, the rates have in practice been determined rather arbitrarily. In addition, over time, various rebates and reductions, motivated primarily by equity considerations, have extended to specified polluters and users, thus reducing the overall efficiency of incentives provided by the fees.

At current fee levels, collected revenues in Poland correspond to more than 40 per cent of the country's substantial environmental investment expenditures. This constitutes a reasonable public share in the effort. Hypothetically, increasing the fees to their incentive level would result, in the short run, in revenues much larger that could be efficiently spent on environmental protection. Thus, switching from the current moderate rates to higher incentive rates might reduce or eliminate the scope of earmarking in the future. This reform strategically should correspond to a future shift in tax policy from labor and income taxes to environment-related taxation. However, there is no indication that Poland's parliament will take up that kind of tax reform in coming years.

It must be emphasized that not all polluters face the same price for the same pollution. For reasons of equity municipalities pay reduced rates for waste water discharges and water use. They also have easier access to subsidies provided by environmental funds, and they receive financing on softer terms than industrial polluters. Thus, Polish environmental fees act as a mechanism transferring resources from the industrial to the municipal sector. Fees levied on producers usually have a more distorting effect throughout the whole economy than fees levied on final consumers. Moreover, it sends wrong signal to consumers, who do not change their polluting and resource use behavior, and to industrial plants which look for cheaper opportunities for disposing of their effluents (for instance through municipal waste water treatment plants).

To the extent that environmental fees are subtracted from taxable income and most of their revenues are allocated for environmental policy purposes in the form of subsidies, they can be viewed as an indirect state budgetary contribution to environmental investment programs. It is not clear, however, whether Polish emission fees are intended to play an incentive- or revenue-raising role. As a result, the fees do not do either properly. There is a trade-off between the two roles; the design of a revenue-raising fee should be different from the design of an incentive fee. Currently, it is not clear whether (and, if so, which) environmental fees will gradually evolve into reliable vehicles to raise revenues or into a credible instruments to bring polluters into compliance with the emission standards. As revenue raising devices, current fees are too complex and too expensive to administer for them to be cost-effective. As incentive instruments, fees are not effective because fee rates are well below the marginal costs of abatement and because enforcement (collection) procedures are not structured to handle situations involving really strong incentive effects. When fees severely affect their profits, firms often respond by refusing to pay rather than by investing in abatement or by closing-down. The social problems that have arisen during the transition period, along with difficulties financing environmental investments, increase the incentives for fee-evasion. The incentive role of emission fees is realized mainly through the use of fee revenues to subsidize environmental investments. However, for such a fee-cum-subsidy scheme to be effective, the fees must provide a stable revenue base for the environmental funds. And for this purpose, the OECD experience shows that product fees often are more effective instruments than emission fees for raising revenues.

Non-compliance fines, permanently present in the system, often are praised as an additional revenue raising mechanism. However, this attitude toward non-compliance undermines the role of fines as an instrument designed to penalize excessive pollution. Moreover, the execution and collection of non-compliance fines in Poland is very poor. An alternative strategy should limit the application of non-compliance fines to incidental spill-overs, and exclude perpetual fine payments by chronic polluters. In place of the non-compliance fines, chronic violators would be disciplined by direct regulation. Economists and lawyers agree with this approach, pointing out that when penalty payments are treated just like regular emissions fees, they undermine the credibility of environmental permits and emissions limitations as instruments designed to prevent unacceptable deterioration of environmental quality.

Polluters in the most industrialized and polluted regions of the country protested effectively in 1992 against regional differentiations in non-compliance fines. This attitude, even if partially supported by arguments of constitutional equity, is not acceptable on ecological or economic grounds.

Product fees presently are a matter of political discussion and research. A recent report prepared for the Ministry of Environmental Protection offers recommendations for introduction of product fees:

- differentiate surcharge levels depending on fuels;
- introduction of a 6 per cent maximum level, predominantly for high-sulphur content coal and high-lead content gasoline;
- introduction of a zero per cent surcharge level for least noxious fuels, including coal and oil of low sulphur content or unleaded gasoline; and
- endorsement of intermediate rate levels selected from within the zero to 6 per cent range for other fuels.

The suggested rates for environmental surcharges on fuels are below levels that would compel economic agents to meet regulatory requirements for emissions. These rates recognize implicitly that it would be politically infeasible to introduce fees at levels that induce the necessary emissions reductions. It should be stressed, however, that even relatively low surcharges should motivate users to curb emissions, in particular by changing the structure of fuel consumption.

In principle, the Polish permitting schemes, with their ambient concentration ceilings and individual emission quotas, provide a firm foundation for a system of transferable pollution quotas. Indeed, emissions trading seems to be a logical extension of current permitting procedures. However, while there has been some interest in implementing this instrument, only one attempt has been made to put emissions trading into practice in Poland, and that application was significantly different from a real market of emission permits. Nevertheless, some interesting results were achieved in that so-called 'Chorzów experiment' (Dudek et al., 1992). Despite numerous controversies and debates, tradeable permits have become a core element of American economic regulations in air pollution policies dealing with sulfur dioxide and nitrogen oxide emissions. Poland has an opportunity to be in the forefront of European countries which are just beginning to apply this promising instrument. However, Poland's environmental laws must be amended to provide more flexibility in such permitting schemes.

In spite of the progress made in applying and enforcing economic instruments for environmental protection, Poland's present system of fees still is far from perfect. On one hand, it is plagued by regulatory irrationalities inherited from the old system. On the other hand, new political problems, such as the tendency towards administrative decentralization, may influence dramatically fee collections and the financing of environmental protection. A few problems illustrate the weaknesses in Poland's present system of economic regulation (Żylicz, 1994):

1. Given Poland's monitoring capacity, the fee lists, especially those for air pollutants, are excessive. For many air pollutants there is no approved methodology for measuring or monitoring emissions.

2. The differentials in fee rates across various pollutants are not necessarily justified by their marginal damages. Many of these rates are based on toxicity, but they are not necessarily correlated with economic measures of damages.

3. Efficiency is impaired by equity measures incorporated into the system, especially in the case of water pollution fees. For example, fees for the use of surface and ground waters are much higher for certain classes of industrial users than for households, municipalities, and agricultural interests. Fees on waste water discharges display similar distortions, although the range of fees is less extreme.

4. Additional distortions result from the method by which fees are recirculated for environmental investments. The most efficient method of distributing fee revenues would be to allocate them to projects generating the greatest net social benefits. However, for three major categories of fees — saline water discharges from coal mining, sulfur dioxide emissions, and nitrogen oxide emissions — revenues are earmarked for investments in curbing the pollution from which those fees are derived.

5. Severe distortions in the efficient use of fee revenues has resulted from the fact that, in net terms, Poland's environmental funds have acted as agencies transferring resources from manufacturing industries to municipalities. This practice reinforces and, in the case of water fees, amplifies the wrong signals sent to the household sector and municipalities.

6. The National Environmental Protection Fund and provincial funds have employed poor project screening and assessment procedures in evaluating grant and loan applications. These practices sometimes lead to the selection of projects that are not cost-effective, even within their respective categories of abatement.

7. A general decline in fee and fine collection efficiency is evident. From 1990 to 1993 the amount of imposed fees and fines increased by factor of 21. Revenues, on the other hand, increased only 14-fold during the same period. The revenues collected have not kept pace with the amounts imposed because of the financial hardship experienced by many economic entities and sectors during the reconstruction of the national economy.

8. In addition, pressure from local governments has resulted in proposals to further decentralize the fees. In fact, fees already have been decentralized to a large extent, with the creation of provincial funds and, since mid-1993, incipient municipal environmental funds. The benefits of further decentralization are questionable; more localized control may reduce the effectiveness of environmental investments, and the dispersion of funds could extend the average construction time of projects. Moreover, national environmental priorities are less likely to receive adequate support. Nevertheless, arguments in favor of returning pollution fees to local projects where the pollution actually takes place are difficult to rebut in the absence of more analysis and a stronger lobby for regional and national priorities.

9. The resistance to product fees, which are viewed as another potential source of revenues for the National Environmental Fund, is inconsistent with the widespread expectation that 100 per cent of emissions fees and non-compliance fines should be retained at the provincial or even municipal level. Indeed, to the extent that some national funding remains desirable, the only way to reconcile the two propositions is to accept the idea of product fees. Product fees levied on environmentally harmful goods could not be convincingly claimed by any single provincial or municipal administration, and would provide a natural source of revenues for the national fund. That would permit the National Fund to survive even if all emissions and resource use fees remained at the provincial and municipal levels.

Recommendations

Poland's system of environmental fees is more than 15 years old. Its effectiveness has improved since 1990 as fee rates have increased and market reforms have taken hold. The fee levels are high by OECD standards, but they are still below Pigouvian levels (Anderson and Fiedor, 1995). Their positive impact is largely in providing revenues for the special environmental funds which have played a key role in boosting environmental protection investment expenditures since 1990.

General Recommendations

The following five general recommendations for improving the current fee system concern priorities likely to enhance the efficiency of Poland's environmental policy, including the performance of economic instruments, without compromising its apparent short-term effectiveness (Żylicz, 1994; Anderson and Fiedor, 1995):

- *Fine-tune fee rates*. The system of fees should be streamlined and simplified to enhance monitoring and collection, and absolute and differentiated fee rates should be corrected.
- *Combine decentralization of emission fees with the introduction of product fees*. Revenue from product fees would go to the National Fund, allowing some reallocation of emission and resource use fee revenues from the National Fund to provincial and municipal funds.
- *Increase flexibility in permitting*. The introduction of marketable permits as an option to achieve cost-effectiveness in solving local problems related to the environment, restructuring, development, and employment.
- *Clarify responsibilities for meeting ambient quality standards among different levels of government*. In particular, authorization of seven regional water agencies delineated according to the major river basins in the country.
- *Scrutinize operating procedures and project selection procedures of the National Environmental Fund to promote cost-effectiveness in environmental investments*.

On one hand, the expansion of traditional economic incentives now seems to have reached its limit, especially as concerns the scope and levels of emission fees. On the other hand, there is a critical need to improve the environmental performance of many heavy polluters, some of which are in chronic non-compliance with current regulations. This points up the continuing need for non-tax instruments to achieve short-run environmental improvements.

This is where, among other economic measures, transferable pollution permits can play a positive role in addressing inherited sectoral and regional problems in a politically feasible, cost-effective way. The American bubble policy (according to which a number of discrete pollution sources owned by one company, such as individual smokestacks, are placed under a imaginary bubble and treated as a single source, permitting the owner more flexibility in achieving required emissions reductions) and the Polish experiment at Chorzów are useful starting points for further studies and effective implementation of emissions trading schemes. However, under Poland's current regulatory regime, emissions limitations (which are, in essence, entitlements or quotas) are specificized for each individual source, and plants have no authority to dispose of or transfer entitlements to other sources. The introduction of an emissions trading policy will, therefore, require changes in the legal rules concerning the setting of emissions limits. Administrators must be authorized to take a differential approach to pollution sources, and permit more flexibility in the use of emissions allowances (Managing the Environment, 1994).

The list of recommendations set forth below includes more specific suggestions on how to improve the use of economic instruments, to harness market forces in the service of environmental policy (Śleszyński, 1996):

Recommendations for Pollution and Resource Use Fees

- Clarify the authority of Water Basin Agencies to collect fees on water withdrawal and water pollution, thus giving them full responsibility for water quality and water supply. This move obviously must be strictly correlated with the introduction of product fees, in order to provide supplemental funding sources for the National Fund and provincial environmental funds.
- Verify lists of pollutants for which fees apply, and cancel emission fees which have only a marginal influence on pollution levels and revenues (if necessary, identify hazardous pollutants to be assessed at higher rates to compensate for any revenue losses).
- Decrease the gap between the emission fees charged to industrial pollution sources and the lower fees charged to low emissions sources, municipalitics and households.
- Clarify as much as possible the distinction between economic instruments designed to raise revenues and those designed to induce pollution abatement. If a fee is intended to provide an incentive effect, it should be set according to the analytically identified marginal abatement costs and introduced carefully in a step-by-step manner. Revenue raising fees, by contrast, should be designed to minimize

administrative costs, for example by imposing flat rates and simplifying measurement and collection procedures.

- Allow for regional differentiations of emission fees and non-compliance fines *and* give provincial authorities the legal responsibility for ambient environmental quality. In principle this should increase their flexibility in regulating individual pollution sources. However, there should not be too much administrative discretion without prior strengthening of environmental control and administration.
- Strengthen mechanisms to ensure fair revaluations of environmental fees with regard to anticipated inflation.
- Offer new market-based solutions (such as performance bonds and environmental liability schemes) as supplements to direct regulation, especially for firms in chronic non-compliance.
- Change the present rule limiting the use of fees collected for employing agricultural land for non-agricultural purposes. Currently, those fees cannot be used to recultivate or reclaim contaminated lands. Eventually, three-quarters of this money could be used for recultivation or reclamation of lands deteriorated by industry.
- Introduce product fees, especially differentiated fuel fees, as stable revenue-raising mechanisms for environmental funds, particularly the National Fund. Products charges should be selected to reduce administrative costs, raise revenues, and control dispersed and mobile sources of pollution.
- Develop a system of deposits for packages, batteries, pesticides and hazardous waste. These deposits should limit stream of waste flowing into landfills.
- Introduce transferable discharge permits as an alternative to increasing emission fees. An emissions trading scheme may be a more efficient economic solution for currently non-complying polluters. In particular, it is a useful vehicle for achieving emissions reductions in the energy sector until 2004, and across all sectors in regions where environmental improvements should minimize associated social costs. The introduction of emissions trading should begin with the internal netting of emissions from multiple sources at individual plants and the bilateral exchange of permits, and then develop into a regional bubble policy.
- Establish a reliable monitoring system for CO_2 and methane, then raise fees for emissions of those greenhouse gases to comply with international legal obligations.

- Transform economic instruments, in the long run, to perform more as a tools of prevention rather than end-of-pipe emissions reduction mechanisms.
- Support the further development of the national system for collecting information and statistics on environmental expenditures. In particular, establish pilot projects, in cooperation with the Main Statistical Office, to collect data on the environmental costs incurred by local authorities, enterprises, and households. This would enable more appropriate analyses of environmental policy performance.

Recommendations for Non-compliance Fines

- Limit the role of non-compliance fines to penalize incidental pollution in excess of permitted standards, instead of tolerating perpetual collections of fines from sources chronically in non-compliance.

Recommendations for Other Taxes and Duties

- Enrich economic regulations through more significant tax differentiations, tax and tariff allowances, and the establishment of escrow funds to promote environmentally benign production, activities, and investments.
- Elaborate a strategy for future tax reform which would shift financial burdens from labor and capital to the environmentally harmful activities of producers and consumers.

Recommendations for Subsidies

- Simplify collection rules and related financial flows between polluters and numerous environmental funds, especially for NO_x.
- Continue subsidizing environmental investments through the environmental funds, but improve their procedures for project evaluation and selection.
- Allow for the further development of provincial and municipal environmental funds according to their individual needs and potential, but prevent further dispersions of financial funds (for instance through industry-based environmental funds).

Conclusions

A series of significant rate increases (supplemented with a new system of more frequent collection) following 1989 improved the operation of Poland's system of ecological charges (Smith and Żylicz, 1994). Despite the economic recession that followed on the heels of 'shock therapy' economic reforms in 1990-1992, Poland's environmental fee revenues are now 15-20 times higher in real terms than at the beginning of 1990. As a matter of fact, even during the first years of economic transformation, Poland's pollution fees amounted to almost 0.5 per cent of the country's GDP (Anderson and Żylicz, 1993). To a certain extent, economic transformation itself has facilitated environmental improvements, and environmental investment effects indisputably have contributed to the general environmental improvements.

However, a long list of environmental policy performance problems still needs to be addressed and eliminated. Very few studies, for instance, have undertaken comprehensive macroeconomic analyses of the performance of economic instruments. The abatement incentive role of economic instruments is highly questionable. On the other hand, the role of pollution and resource use fees as revenue-raising mechanisms is well-understood and acknowledged.

In one study based on a sample of 112 major industrial polluters, the authors surveyed emissions, abatement rates, pollution fees, abatement costs, types of abatement projects, investment expenditures and their environmental effects (Broniewicz et al., 1994). This sample provides basic information for analyzing polluters' responses to environmental policy instruments. It was estimated that in 1992 their air pollution emissions and waste water discharges were responsible for damages on the order of US$ 500 million. At the same time, the enterprises paid US$ 105 million in air and water pollution fees. The recurrent costs of their abatement activities were US$ 84 million. In addition, they spent US$ 108 on environmental investment projects. Those investment expenditures were financed mainly from polluters' profits (58 per cent), commercial credits (22 per cent), and soft loans or subsidies from environmental funds (recirculating pollution fees) (16 per cent); contributions from the state budget and foreign assistance were negligible (less than 4 per cent). What is obvious is the gap between the value of damages and the cost of abatement effort, and the fact that the sample plants received back in the form of subsidies and soft loans only a small portion of pollution fees they paid. This comparison speaks for itself, but it was only a sample investigation.

It seems rather obvious that environmental funds have proven to be an important element of environmental financing. Their activity is enabled by the significant fee and fine collections, which they turn around and distribute in the form of subsidies and soft loans. Indeed, Poland's environmental funds provide useful models for other countries in transition, especially for those which still have problems securing budget monies for environmental purposes

(Śleszyński, 1996). The transition period has not been one in which large environmental programs have flourished, but it is still possible to attain quite satisfying levels of environmental investment, depending on the effective coordination of social environmental awareness, political will, and institutional and regulatory structures. The significance of Poland's experiences for developed countries will become clear over time, as Poland's environmental regulations are harmonized with European Union Directives and, ultimately, as Poland (and other transitional economies) join the EU. At that point, the question will be whether Poland's sophisticated and expensive system of environmental fees will be viewed as a curious phenomenon without broader applicability, or will it serve as a model for general directives on environmental pricing. It would be ironic indeed if Poland's system of economic instruments for environmental protection, which are largely an inheritance from the socialist system, changed the face of environmental regulation in Western Europe. And it would be just as ironic if Poland's market mechanisms for environmental protection vanished step-by-step along the road to a unified European environmental policy.

The appropriate solution should be in line with long-run strategies for economic regulation and taxation, as first implemented by the Scandinavian countries. It should also rely on reasonable decisions concerning preferences for specific new economic instruments (including product fees and tradeable emission permits) and eliminate ineffective or inefficient regulations. All in all, it is very likely that Poland's environmental policy for the rest of this century will continue to rely on the old fee system, while developing new economic instruments and adapting EU regulations.

Notes

1. In 1995 Poland replaced its old currency. One new zloty was equivalent to 10,000 old ones. As of January 1996, 2.55 new zlotys equaled one U.S. dollar.

2. In July 1993 Poland's sales tax, which differentiated among product and service categories, was replaced by a value-added tax (VAT). At present, the regular VAT rate is 22 per cent, and the special rate (for specific cases) is seven per cent.

Annex

Tables on environmental and resource fee rates, revenue raised, and fee collection efficiency 1990-1996.

Table 3.A1 Emission Fees for Sulphur Dioxide

Rates:	1990	1.89 next 28.42 US$/ton	
	1991	64.25 US$/ton	
	1992	59.49 US$/ton	
	1993	42.44 next 66.13 US$/ton	
	1994	66.01 US$/ton	
	1995	78.37 US$/ton	
	1996	94.08 US$/ton	
Revenues:	1990	7.05 million US$	(99% imposed)
	1991	121.31 million US$	(86% imposed)
	1992	111.47 million US$	(86% imposed)
	1993	186.10 million US$	(150% imposed)
	1994	105.26 million US$	(97% imposed)

Non-compliance fine is 10 times the regular fee rate.

Source: Spyrka, 1996.

Table 3.A2 Emission Fees for Nitrogen Oxides

Rates:	1990	0.51 next 28.42 US$/ton	
	1991	64.25 US$/ton	
	1992	56.49 US$/ton	
	1993	42.44 next 66.13 US$/ton	
	1994	66.01 US$/ton	
	1995	78.37 US$/ton	
	1996	94.08 US$/ton	
Revenues:	1990	0.67 million US$	(94% imposed)
	1991	32.16 million US$	(85% imposed)
	1992	33.27 million US$	(80% imposed)
	1993	35.74 million US$	(84% imposed)
	1994	36.69 million US$	(96% imposed)

Non-compliance fine is 10 times the regular fee rate.

Source: Spyrka, 1996.

Table 3.A3 Pollution Fees for Waste Water

Biochemical oxygen demand during the first 5 days, rates:

1990	11.16 next 167.37 US$/ton
1991	524.43 US$/ton
1992	712.71 US$/ton
1993	535.41 US$/ton
1994	704.10 US$/ton
1995	816.70 US$/ton
1996	991.77 US$/ton

Chemical oxygen demand, rates:

1990	6.74 next 101.05 US$/ton
1991	299.54 US$/ton
1992	407.16 US$/ton
1993	305.87 US$/ton
1994	396.06 US$/ton
1995	457.85 US$/ton
1996	556.64 US$/ton

Suspended solids, rates:

1990	1.05 next 15.79 US$/ton
1991	46.79 US$/ton
1992	63.83 US$/ton
1993	47.95 US$/ton
1994	61.61 US$/ton
1995	70.12 US$/ton
1996	85.06 US$/ton

Heavy metals (total mass), rates:

1992	7,336.22 US$/ton
1993	5,511.16 US$/ton
1994	7,041.01 US$/ton
1995	8,158.72 US$/ton
1996	9,917.68 US$/ton

Chlorate and sulphate ions, rates:

1990	0.53 next 7.89 US$/ton
1991	23.62 US$/ton
1992	40.35 US$/ton
1993	30.31 US$/ton
1994	39.61 US$/ton
1995	45.37 US$/ton
1996	55.15 US$/ton

Total revenues:

1990	6.71 million US$	(108% imposed)
1991	107.05 million US$	(69% imposed)
1992	104.43 million US$	(50% imposed)
1993	74.82 million US$	(55% imposed)
1994	91.30 million US$	(45% imposed)

Non-compliance fines are fixed numbers.

Source: Spyrka, 1996.

Table 3.A4 Pollution Fees for Saline Water (Coal Mining)

Chlorate and sulphate ions content, rates:

1990	0.53 next 7.89 US$/ton
1991	23.62 US$/ton
1992	40.35 US$/ton
1993	30.31 US$/ton
1994	39.61 US$/ton
1995	45.37 US$/ton
1996	55.15 US$/ton

Revenues:

1990	2.02 million US$	(119% imposed)
1991	38.27 million US$	(62% imposed)
1992	5.19 million US$	(6% imposed)
1993	2.05 million US$	(5% imposed)
1994	14.02 million US$	(12% imposed)

Non-compliance fine rate for saline (coal mining) water is a fixed number.

Source: Spyrka, 1996.

Table 3.A5 Fees for Surface and Ground Water Withdrawal

Surface water withdrawal, basic rates:

1990	0.51 next 7.58 US$/thousand m^3
1991	17.01 US$/thousand m^3
1992	23.11 US$/thousand m^3
1993	17.36 US$/thousand m^3
1994	22.00 US$/thousand m^3
1995	24.75 US$/thousand m^3
1996	30.18 US$/thousand m^3

Ground water withdrawals, basic rates:

1990	1.52 next 22.74 US$/thousand m^3
1991	51.03 US$/thousand m^3
1992	69.33 US$/thousand m^3
1993	52.08 US$/thousand m^3
1994	66.01 US$/thousand m^3
1995	78.37 US$/thousand m^3
1996	95.26 US$/thousand m^3

Total revenues:

1990	2.71 million US$		(105% imposed)
1991	38.38 million US$	(77% imposed)	
1992	54.92 million US$	(83% imposed)	
1993	41.60 million US$	(75% imposed)	
1994	54.64 million US$	(96% imposed)	

Source: Spyrka, 1996.

Table 3.A6 Fees for Waste Disposal

Range of differentiated rates:

1990	0.01-0.51 next 0.19- 7.58 US$/ton
1991	0.47-17.01 US$/ton
1992	0.73-14.67 US$/ton
1993	0.55-11.02 next 1.10-16.53 US$/ton
1994	1.32-17.60 US$/ton
1995	1.53-20.39 US$/ton
1996	1.86-24.79 US$/ton

Revenues:

1990	3.83 million US$	(107% imposed)
1991	43.08 million US$	(63% imposed)
1992	55.87 million US$	(86% imposed)
1993	62.92 million US$	(68% imposed)
1994	63.54 million US$	(78% imposed)

Non-compliance fines on waste disposal are related to the regular fees. They are assessed as 1/20 of the individual, annual fee per ton per every day of depositing the waste without a valid permit.

Source: Spyrka, 1996.

References

Anderson, G. and B. Fiedor (1995), Environmental Fees in Poland. Paper presented at conferences in Sofia and Vilnius.

Anderson, G. and T. Żylicz (1993), Environmental Policy in Poland: Prospects for Economic Instruments. Paper Presented at the Seminar on Industrial Pollution Permitting in Central and Eastern Europe, Vilnius.

Broniewicz, E., Poskrobko, B. and T. Żylicz (1994), Internalizing environmental impacts of industry in Poland: preliminary empirical evidence. Paper presented at EAERE Conference, Umea.

Case Study of Environmental Expenditure and Investment in Six Selected CEE Countries. Draft Final Report. OECD/COWIconsult, 1995.

Dudek, D.J., Kulczyński, Z., and T. Żylicz (1992), Implementing Tradeable Pollution Rights in Poland. Paper presented at the Third Annual Conference of the EAERE, Kraków.

Informacja o wymiarze I wpływach z opłat za gospodarcze korzystanie ze środowiska I kar za nieprzestrzeganie wymogów jego ochrony oraz o rozdziale tych środków realizowanych w roku 1994 przez wojewodów, 1994, Narodowy

Fundusz Ochrony Środowiska I Gospodarki Wodnej, Warszawa. [Report of the National Fund of Environmental Protection and Water Management]

Informacje na temat źródeł I sposobów finansowania inwestycji z zakresu ochrony środowiska oraz uzyskanych efektów w latach 1991-1993, 1994. Departament Polityki Ekologicznej, MOŚZNiL, Warszawa. [Information on sources and methods of environmental financing provided by the Ministry of Environmental Protection]

Kruszewski, J. (1994), 'Poland', in Francis, P. (ed.), *National Environmental Protection Funds in Central and Eastern Europe. Case studies of Bulgaria, the Czech Republic, Hungary, Poland and the Slovak Republic*, Regional Environmental Center, Budapest, pp. 56-70.

Managing the Environment. The Role of Economic Instruments, 1994. OECD, Paris.

National Environmental Policy. Outline of Economic Instruments, Ministry of Environment, Warsaw, 1990.

Ochrona środowiska 1994, 1994, GUS, Warszaw. [Environmental protection yearbook 1994]

Ochrona środowiska 1995, 1995, GUS, Warszawa. [Environmental protection yearbook 1995]

Smith, S. and T. Żylicz (1994), 'Workshop on Taxation and Environment in European Economies in Transition: Rapporteur's Report of the Discussion and Conclusions', in *Taxation and Environment in European Economies in Transition*. Paris, OECD, 6-15.

Spyrka, J. (1994), 'Poland. Part B', in J. Klarer (ed), *Use of Economic Instruments in Environmental Protection in Central and Eastern Europe. Case Studies of Bulgaria, the Czech Republic, Hungary, Poland, Romania, the Slovak Republic, and Slovenia*, Regional Environmental Center, Budapest, 105-121.

Spyrka J. (1996), 'Zasady funkcjonowania instrumentów ochrony środowiska w Polsce', Warsaw University, Economics Department, Warsaw Ecological Economics Center, Warsaw, to be published. [Foundations of the functioning of environmental protection economic instruments in Poland]

Śleszyński, J. (1996), 'Economic instruments in the environmental policy: country profile Poland', forthcoming in *European Environment.*

Żylicz, T (1994), 'Taxation and Environment in Poland', in *Taxation and Environment in European Economies in Transition,* OECD, Paris, 36-56.

Żylicz, T. (1996), 'Economic instruments in Poland's environmental policy', Warsaw University, Economics Department, Warsaw Ecological Economics Center, Warsaw.

Żylicz, T. and Z. Lehoczki (1994), 'Towards Environmental Recovery. The Czech Republic - Hungary - Poland - Slovakia', in Domański R. and E. Judge (eds), *Changes in the Regional Economy in the Period of System Transformation*, PWN, Warszawa, pp. 91-154.

4 Privatizing the Polish Economy: Benefits and Dangers to the Natural Environment

BOGUSŁAW FIEDOR

Introduction: Problems Identifying the Ecological Effects of Privatization

This paper is not intended to be a comprehensive analysis of the impact of the privatization process on Poland's natural environment; it merely outlines the most important ecological consequences and raises certain issues that should be addressed in future theoretical and practical studies.

Research on the ecological aspects of privatization in Poland, or in any other economy making the transition from a command-and-control system to a market economy, should be based on the following theses:

First, the ecological effects of privatization cannot be 'isolated' from the effects of other aspects of the transformation, which include: macroeconomic stabilization; de-monopolization; the substantial reform of such basic macroeconomic areas as price, tax, and commercial policies; and the development of industrial, energy, transportation, and agricultural policies.

Second, research into the relation between privatization and the natural environment must refer back to the systemic causes of environmental degradation under the old command-and-control system and Poland's specific patterns of economic development and industrialization.

Third, at the present stage of economic transformation in Poland, the vast majority of enterprises, regardless of their ownership status, do not obey the rules described in standard microeconomic textbooks. Therefore, the environmental impacts generated by the privatization process are bound to be quite different from those generated by private enterprises operating in mature competitive markets, where scarce resources are allocated by price and where there is a potential for technological substitution.

Fourth, the ecological impacts of privatization may differ in the long- and short-run. In the long-run, the general increase in economic efficiency that is expected to result from the commercialization and privatization of state-owned enterprises may produce significant ecological benefits. But in the short-run, during the period of privatization proper, the natural environment may suffer, for instance because of the lack of adequate environmental controls on the privatization process (See *Schedule of actions for natural environmental protection in Central and Eastern Europe*, 1994, pp. 23-39).

Fifth, many liberal economists and politicians mistakenly believe that because the socialist economy caused an advanced degradation of the environment in Poland, the transition to the market economy will automatically solve the country's basic ecological problems. This assumption may prove harmful to the development of a sound ecological policy. It may be compared to the misguided notion that market economies do not require industrial or energy policies. They do, of course, because even market economies based on private property experience market failures which impede environmental protection and the conservation of natural resources. The market mechanism has proven incapable, by itself, of achieving or maintaining an ecologically sustainable level of development. This is clear from the considerable environmental degradation that all private property-based market economies have suffered in the process of their development. Thus, while privatization may bring certain ecological benefits (many of which will be discussed below), it will not, by itself, improve the quality of the natural environment in Poland. It must be accompanied by an environmental policy that addresses the ecological externalities resulting from market failures.

Economic Transformation vs. the Natural Environment in Poland

Two main sets of factors contributed to environmental degradation in socialist Poland: first, the specific character of the socialist command-and-control economy and the pattern of economic development in Poland; and second, the inability of narrowly conceived environmental policies to counter the ecologically damaging effects of Poland's economic development.

Among the most important factors related to Poland's socialist economic development were:

- Poland's autarkic model of development resulted in technological backwardness and an ecologically disastrous pattern of primary and final energy consumption.
- The pattern of industrialization, based on the dominance of heavy industry and the raw material and fuel-energy sectors, led to an ecologically damaging structure of output.
- Poland's accelerated capital formation, with its focus on developing the fuel-energy and metallurgical sectors, went far beyond the

requirements of balanced economic growth.

- Mainly on ideological grounds, Poland's economic links with the Council of Mutual Economic Assistance (the defunct trade organization of former Soviet Bloc countries) were based on the assumption of self-sufficiency of raw materials and energy, and on a supply-oriented development strategy in the fuel-energy sector.
- The politically dominant position of state-owned enterprises rendered legal-administrative and economic tools of environmental policy ineffective.
- The reward structure of the system of central economic planning, which valued quantitative and physical measures over financial measures of enterprise performance, resulted in excessive and often wasteful exploitation of natural resources.
- As a matter of policy, the prices of energy, raw materials, and natural resources was kept below the costs of their production, extraction, transmission, and distribution. This led directly to ecological degradation through the excessive consumption and depletion of these resources.
- Rapid urbanization was connected with the location of new towns and housing districts very close to large industrial complexes, which exposed a large part of the population to the worst environmental hazards.
- Socialism's endemic soft budget constraints rendered state-owned enterprises insensitive to cost and price signals, leading them to (rationally) overuse energy and raw materials and to pollute excessively.[1]

Of these various factors, the most important was the soft budget constraint on state-owned enterprises. Since enterprises did not have to profit to survive, it did not matter that their production costs greatly exceeded the value of what they produced. Consequently, the energy- and raw material-intensity of production in Poland's socialist economy was two to four times higher than the OECD average (see, for example, Żylicz, 1992). The inability to improve productive efficiency in Poland was primarily responsible for the country's (absolutely and comparatively) high levels of pollution and resource overuse.

The most important factors related to socialist Poland's inability to adopt and implement effective environmental policies included the following:

- Priorities for environmental protection were never clearly formulated.
- The system of environmental management was not adequately

decentralized, and regional authorities lacked sufficient powers.

- The 'shortage economy' and the soft budget constraint of state-owned enterprises rendered economic instruments of environmental protection ineffective.
- The state's overlapping and conflicting functions as owner and regulator of polluting enterprises weakened the enforcement of environmental standards and other instruments of direct regulation.

Given that the centrally managed socialist economy proved to be both economically inefficient and incapable of meeting the broadly conceived ecological objectives associated with modern development, it is natural to assume that the transition to a market economy should bring ecological as well as economic benefits. This belief, which is encountered both in the popular press and less often in research papers dealing with ecological and economic issues, is based on the fact that during the 1970s and 1980s most of the industrial market economies in the OECD achieved outstanding results cleaning up the natural environment and rationalizing the consumption of natural resources. As usual, however, the social and economic realities are more complicated than they appear in this portrayal.

The continued presence of market failures relating to environmental protection and natural resource use in advanced market economies clearly demonstrates the need for a well-designed and consistent environmental policy. Such a policy implicitly presupposes that the state possesses the political will to protect and restore the natural environment. It also requires a high level of ecological awareness among both producers and consumers, to transform a general societal desire for a cleaner environment into the actual willingness to pay for expensive environmental programs.

This brief discussion of the social and institutional premises of environmental protection is not meant to suggest that there are no potential sources of environmental improvement directly related to the transformation of the Polish economy. The privatization of state-owned enterprises, in particular, can lead to improved environmental quality in at least two ways, both related to the hardening of budget constraints. First, hardened budget constraints should motivate firms to improve productive efficiency, leading to a reduction in the economy's energy- and raw material-intensity. The potentially positive effects of this reduction are particularly important given Poland's extremely damaging past patterns of energy and resource use. In addition, hardened budget constraints should render firms more sensitive to price signals, including the fees and fines of Poland's market-based environmental policies. However, even these beneficial ecological consequences of privatization and marketization will not come automatically or unconditionally. They depend on many other factors, the most important of which are as follows:

- The government must consistently implement environmental policies and enforce environmental laws.
- Market-oriented incentives to reduce the intensity of energy and raw materials use must be strengthened by a coherent energy and industrial policy, which includes the dissemination of information, research and development policy, preferential tax allowance, and so on.
- The goals and instruments of environmental policy must be harmonized with privatization, energy, and industrial policies, as well as with governmental policies toward economic sectors such as agriculture and transportation.

Privatization and the Environmental Reconstruction of the Polish Economy

Potential Positive Effects

Since 1990 the levels of environmental fees and fines charged for pollution emissions and waste disposal in Poland were increased about 10 times in real terms. For many economic sectors and industrial branches, this caused a substantial rise in environmental protection costs as a percentage of total production costs. Assuming that producers are more sensitive to changes in relative prices (because they face harder budget constraints) and that a genuine demand constraint has appeared, the opportunities to transfer increased costs of environmental protection to final product prices should be reduced. Thus, increasing the levels of environmental fees and fines should provide greater incentive for firms to abate emissions and effluents, either by adopting environmentally-friendly technology or by modifying the structure of the economy's final product.

Under socialism, state-owned enterprises tended not to be motivated by instruments of indirect regulation or indirect public subsidies, such as tax allowances or preferential loans for environmental investment projects. But in post-Communist Poland, the future diffusion of 'least-cost' or profit-oriented strategies for privatized enterprises should significantly increase their incentive to abate pollution and reduce resource waste. It should be noted that although the use of preferential loans or tax relief for financing ecologically friendly activities in private firms has been roundly criticized in Poland and in some OECD countries, these methods have proven beneficial for achieving the environmental protection policy objectives in countries such as Germany and

Sweden.

The privatization of state-owned enterprises and, more generally, the increasing contribution of the private sector to growth and production also has important implications for the effectiveness of environmental policies in Poland. Specifically, the functions and interests of those who now own the means of production and other economic resources are being steadily separated from the functions and interests of those responsible for establishing and implementing environmental policies and enforcing regulations. However, the actual significance of this institutional development will depend on the consistency of law enforcement by national and regional governments. The greatest danger for environmental protection lies in the potential conflict between the desire to maintain a stable or growing level of employment (especially at the regional and local levels) and the need to cut production in or close down entirely plants that chronically violate environmental protection laws. This danger may be minimized, however, through the adoption of more flexible regulatory methods and instruments, such as 'escrow funds' and performance bonds (currently in use in the United States), which can assist firms in the process of privatization to meet environmental standards.

As a result of privatization and the gradual de-monopolization of the economy, the widespread diffusion of least-cost operational strategies in Polish enterprises may bring about some beneficial ecological consequences through changes in the sectoral and branch structure of the economy. However, these changes may not be enough to solve the complex problems of environmental protection in Poland. That will require, in addition to environmental policies, active energy and industrial policies modeled on the countries of the European Union and the OECD. In the absence of energy and industrial policies to reduce the energy-, transport-, and resource-intensity of social product to levels comparable to the OECD countries in the 1970s and 1980s, Poland will not be able to reduce the structural burdens on the environment enough to effect a radical clean-up.[2]

Potential Problems and Dangers

The problem of ecological liability. To be efficient a market must be dominated by privately owned enterprises. However, the process of privatization has a great impact on the natural environment both in the long- and the short-run, especially as it relates to *environmental liability*. According to the 'polluter pays principle' (PPP), firms must be liable to compensate for the environmental damages they cause. In countries such as the United States and Germany, where environmental liability is broadly applied, legal regulations define the types and ranges of liability.

In the privatization process, it is necessary to distinguish between liability for current pollution and liability for past contamination. The question of liability for current pollution is uncontroversial, although, as will be discussed below, the application of liability for current pollution in Poland's privatization procedures has made the attainment of certain environmental objectives problematic. Far more controversial is the issue of liability for past contamination, usually involving wastes, contaminated soils, and (more rarely) permanently contaminated surface- and groundwater. Uncertainty over potential liability for past contamination deters domestic and foreign investors — especially foreign investors from countries such as the United States, Germany, and the Netherlands, with extensive environmental liability regimes — from buying the assets of firms undergoing privatization.

The best solution for financing the clean-up of past contamination, it seems, is to cover the costs with public monies from general tax revenues or Poland's various ecological funds, including the National Fund for Environmental Protection and Water Resources Management. This approach would be both ethically appropriate and in accordance with the 'polluter pays principle' (in view of the state's previous ownership of the polluting enterprises). On the other hand, the PPP rules out the use of public funds for the abatement of current pollution, except in specific cases (insolvency, for example).

The potential liability for past environmental contamination also can affect asset valuations of enterprises undergoing privatization. This is particularly important for so-called strategic investors. Including the costs of environmental remediation in the price of enterprises undergoing privatization may be an acceptable alternative to public financing of clean-up. Of course, in order to properly value the environmental costs, it is necessary to conduct detailed ecological audits. Once the costs of clean-up are ascertained, the foreign investor must deposit a sufficient sum of money in a Polish bank to cover the costs of environmental remediation and other environmentally-related aspects of firm restructuring, to be performed in accordance with a time schedule agreed upon in the privatization contract.

Transfer of ecological obligations in different privatization procedures. As mentioned above, one way of dealing with the issue of liability for past contamination is to include costs resulting from the transfer of ecological obligations in the price of the business entity being privatized. In order to ensure environmental protection in the privatization process, regulations governing the transfer of ecological obligations were introduced into all paths of privatization. Researchers from Poland's Institute for Sustainable Development (Stodulski and Starczewska, 1993) have studied the effects of these requirements. They found that:

- Ecological liability tended to be transferred to economically weak firms created from left-over capital.
- Newly founded firms displayed little or no incentive to undertake environmentally friendly investments.
- In complicated ownership transfers, it became difficult to determine who was liable for current pollution and, thus, responsible to pay fees and fines.

The legal succession of environmental liability is particularly difficult to determine when state-owned enterprises are privatized by liquidation. A peculiar 'play for time' often occurs between the founding state organization and business entities that are participating in the process of taking over the liquidated enterprise's capital, leading to a deterioration of the firm's existing environmental protection equipment or to the generation of new pollution (especially solid wastes) as a result of the break in business (see Jakubczyk and Szapiel, 1993).

There is also a serious danger for the transfer of environmental obligations when a state-owned enterprise is liquidated because of its poor financial condition. In such cases, the ecological liability is often assumed by the state because, under Polish bankruptcy laws, the buyer of an insolvent enterprise is not responsible for its debts. It is alarming that the law does not require environmental audits before state-owned enterprises are liquidated, especially since most Polish enterprises privatized through liquidation are in far worse environmental shape than enterprises undergoing capital privatization.

Similarly, there are no uniform legal regulations for conducting environmental audits of enterprises included in the Mass Privatization Program. The Ministry of Environmental Protection, Natural Resources, and Forestry concedes that with more than 200 enterprises involved in that Program, it is very difficult to assess their environmental conditions (OECD, 1995, pp. 92-3). The resulting uncertainty about the environmental status of these firms may obstruct the process of real privatization in the future. In the meantime, it is hardly ethical to deprive investors in privatized enterprises of vital environmental information. The situation is similar in capital privatization, where shares are sold by public offering, but firms are not required to provide environmental information in their prospectuses. (So far, such information has been provided by only a few firms, including Żywiec, Kable, and Polifarb.)

Environmental dangers associated with the creation of small- and medium-sized enterprises. Poland's privatization process has led to the creation of a large number of small- and medium-sized firms. While this has resulted, in some cases, from ownership changes in liquidated state-owned enterprises, most of the expansion of small- and medium-sized enterprises in Poland has been connected to the creation of new private firms.

Two broad sets of potential environmental hazards result from the proliferation of new small- and medium-sized firms in Poland. The first results from the general inadequacy of Poland's ecological monitoring system, particularly that part of the system responsible for measuring emissions rates. The second comes from the haphazardly applied and largely ineffective pollution permitting system, whereby individual waste water discharge, air emissions, and solid waste disposal limits are established for each firm.

Although some measures have been taken in Poland in the last several years to extend and modernize the national and local environmental monitoring networks (often with funding from international financial institutions or from international assistance programs such as the EU's PHARE Program), the system remains relatively ineffective. Thus it is necessary to look for new legal and organizational solutions to facilitate the collection of ecological fines and fees according to the PPP. An example would be the introduction of *lump fees* for small industrial plants and manufacturers, to be assessed on the basis of the types and quantities of energy used, combustion technologies, types and amounts of raw materials used, technologies employed to process them, and so on. Another example would be *voluntary audits* in return for credit preferences to fund the purchase of protective equipment or environmentally-friendly process changes. An *environmental fuel tax* could be set at a level commensurate to the overall ecological harmfulness of a given energy source.[3]

Final Remarks: Privatization For and Against Sustainable Development

Privatization is the most important element of the structural transition from a command-and-control economy to a market economy. But a crucial question is whether privatization encourages or obstructs the transition to a stable and environmentally balanced pattern of development. How does a predominantly private economy relate to the principles of sustainable development, as they have been elaborated in the theoretical literature, the reams of material published by various international organizations, and in official Polish environmental policy?

The table below sketches the direction and strength of influence of privatization processes on the principles of sustainable development:

Table 4.1 Privatization and the Principles of Sustainable Development

	Principle of Sustainable Development	*Nature of the Influence of Privatization*
1.	Abatement of pollution at the source	Positive
2.	Compliance with law	Positive
3.	Common good	Neutral
4.	Economization and the use of market mechanisms	Positive
5.	Polluter-pays principle	Positive
6.	Regionalization	Neutral
7.	Public participation	Neutral
8.	Universality of environmental protection policy and environmental integrity	Neutral
9.	International cooperation	Neutral
10.	Environmental protection policy based on scientific grounds and precautionary principles.	Neutral

Some explanatory points

1. As argued in this paper, privatization does contribute to the abatement of pollution at its source because it stimulates environmentally beneficial structural and technological changes in the economy.

2. Privatization, by legally and formally ending the state's role as owner *and* environmental regulator of polluting enterprises, leads to a greater compliance of producers with the law.

3. The common good may be conceived as inter-regional, inter-generational, or international equity or justice. Privatization cannot be evaluated positively or negatively from the point of view of the special case of ecological justice. But insofar as environmental pollution results from a broad range of factors related to inequity and poverty, privatization can

arguably contribute to sustainable development by increasing economic growth and reducing the degree of inequality and poverty. Moreover, for some private businesses environmentally 'self-limiting' actions may be economically advantageous. For instance an 'ecologically friendly' firm can profit from environmentally-aware consumers.

4. Privatization, in conjunction with deregulation and demonopolization, increases the efficiency of economic instruments in environmental protection because of the increased price sensitivity and concern for productive efficiency. Market instruments also contribute to the efficiency of direct control mechanisms.

5. The 'polluter-pays principle' reflects the idea that the market economy is based on private entrepreneurship. By reducing direct public subsidies of environmental protection, for instance, privatization increases the internalization of ecological externalities by forcing enterprises to consider the negative effects of their activities.

6. It is difficult to predict what tangible influence, positive or negative, privatization will have on the regionalization of environmental protection policy, which may be treated as a special case of subsidy (see, for instance, *Trade and Sustainable Development Principles,* 1994).

7. Public participation is based on 'the principle of openness'. Privatization as such does not cause greater transparency or openness. What matters are legal and institutional conditions that secure the public's right to influence decisions concerning economic activities with significant environmental impacts. As with the principle of the common good, a concern about public relations can motivate owners and boards of directors to pay attention to public participation.

8. All business activities, including those in an economy based on private property and the free market, can exceed the natural limits of ecosystems' regenerative abilities, can cause irreversible changes in biodiversity, and thus can violate the integrity of the environment. For this reason, it is not enough to subject private firms to regulations based exclusively on the 'polluter-pays principle' and the internalization of environmental costs. Direct governmental actions are also necessary, especially to secure inter-generational ecological equity, which is a basic component of the common good principle.

9. When combined with the liberalization of international trade, privatization leads to more diversity and openness in international economic relations. But privatization, in and of itself, does not necessarily directly impact on the methods and intensity of international cooperation to solve interrelated problems of economic growth and environmental protection. Ecological agreements, protocols, conventions, and other international environmental regulations clearly affect the terms of international transactions conducted by private and state-owned Polish enterprises. So, too, do ecological provisions contained in global or regional trade agreements such as the General Agreement on Tariffs and Trade, the World Trade Organization, and the European Union.

10. Privatization has no direct influence on whether or not environmental policy is based on good science.

Notes

1. A more extensive analysis of all the factors underlying the deterioration of the natural environment in Poland is provided in Fiedor et al. (1993).

2. From the beginning of the 1970s to the middle of the 1980s, the OECD countries achieved an increase in per capita GDP of about 30 per cent, while the level of primary energy and end energy consumption barely changed at all.

3. Niedomagała (1994, pp. 224-7) provides interesting statistical data about the problems of environmental protection in small- and medium-sized firms.

References

Fiedor, Bogusław et al. (1993), *Economy-wide Policies and the Environment: A Case Study of Poland*. World Bank, Washington, D.C.

Jakubczyk, Z. and J. Szapiel (1993), 'Ownership transformations vs. environmental protection', paper presented at a conference on 'The Social Market Economy', Sosnowka, Poland, December 3-5, 1993.

Niedomagała, J. (1994), 'Ecological aspects of development of small- and medium-sized firms: Economic system transformation in Poland vs. the natural environment', ESEŚiZN, *Economics and the Environment Library*, no. 16, Wrocław, Poland, pp. 224-227.

Organization for Economic Co-operation and Development. 1995. *Ecological Surveys: Poland*. OECD, Paris-Warsaw.

'Schedule of actions for natural environmental protection in Central and Eastern

Europe' (1994), Ministerial Conference, Lucerne, Switzerland, April 28-30, 1994.
Stodulski, W. and G. Starczewska (1993), *Ownership Transformations in Industry vs. Environmental Protection,* Institute for Sustainable Development, Warsaw.
Trade and Sustainable Development Principles (1994), International Institute for Sustainable Development, Winnipeg.

5 Environmental Law in Poland, 1989-1996: An Assessment of Past Reforms and Future Prospects

JERZY JENDROŚKA

Introduction

Developments in Polish environmental law since 1989 have been far less spectacular and slower than in other Central European countries (Stec, 1994), despite the fact that Poland entered the post-communist era with a legacy of tremendous pollution problems. These problems demanded a new environmental policy, and as with any other policy changes, two questions had to be addressed: What are the goals? And how are these goals to be achieved? The goals were obvious: to combat past, and to prevent future, pollution. Identifying and agreeing upon short-, medium-, and long-term priorities was quite easy. The critical point in setting the agenda, however, was to determine *how* to achieve those priorities.

Many observers anticipated a radical change in Polish environmental law and policy. Exponents of radical change hoped that Poland, and the other countries of Central Europe, would 'leapfrog' the West ecologically by introducing legislation to foster an ecologically sound economy. The existing law would be replaced by a wholly new one employing the most advanced and progressive legal instruments developed in the West. But this has not happened in Poland.

There are many reasons why Poland cannot boast a brand new set of environmental laws. No doubt one reason is that, contrary to common understanding, Poland's prior history of environmental neglect was not caused by the absence of environmental law. Indeed, before 1989 Poland had developed a relatively extensive legislative framework for environmental protection (Radecki and Rotko, 1991, pp. 103-08).

Poland's Environmental Legal Framework Prior to 1989

Roots

The origins of modern environmental law in Poland trace back to the 1922 Water Law, which is considered an early example, on a global time frame, of pollution control legislation (Cole 1995a, p. 300), and the 1935 Nature Protection Act. However, the two most important pieces of pre-1989 environmental legislation in Poland were the 1949 Nature Conservation Act and the 1980 Environmental Protection and Development Act. Their most characteristic feature was a modern, comprehensive approach to environmental protection which distinguished them significantly from piecemeal, media-specific approaches to environmental protection taken, until quite recently, in most countries.

The 1949 Nature Conservation Act has often been characterized as a predecessor to the famous National Environmental Policy Act (NEPA) adopted by the United States in 1969. First, like NEPA, Poland's 1949 Nature Protection Act addressed environmental problems in a comprehensive way; second, it proclaimed environmental protection as a national policy goal; and third, it established a special central agency, akin to the US Council on Environmental Quality, responsible for monitoring compliance with this policy. In addition, the 1949 Act introduced a special action-forcing mechanism that corresponds to NEPA's 'strategic' environmental impact assessment (EIA). But while NEPA became influential worldwide as a model for environmental policy, Poland's 1949 Nature Protection Act remains forgotten. This is because NEPA was enacted and implemented in a political system that facilitated its policies and procedures.

Even though political-economic circumstances prevented Poland's 1949 Nature Protection Act from being implemented, that statute provided a useful background for subsequent legislation. In 1980 Poland enacted the Environmental Protection and Development Act (EPDA), a statute that, in addition to providing basic principles of general environmental policy, also attempted to regulate most pollution problems. The 1980 EPDA was intended as a first step towards establishing an 'environmental code', an idea which recently has gained widespread recognition in western countries,[1] including the United States where the traditional media-specific approach, driven by powerful interests in the legislative branch, has produced a maze of statutes and regulatory powers (Adams and Cox, 1990, p. 10367).

Main Elements

Constitutional Basis. As part of a set of much broader revisions, Poland's constitution was amended in 1976 to include two environmental provisions establishing policy guidelines and citizens' rights:

> Article 12, para. 2: 'The Polish People's Republic ensures the protection and rational development of the environment'.
>
> Article 71: 'Citizens of the Polish People's Republic have the right to enjoy the values of the natural environment and the duty to protect it'.

Pollution Control Scheme. The 1980 EPDA attempted to regulate most pollution problems, including air and soil pollution, noises and vibrations, and waste disposal (see Jastrzębski and Rest, 1982). With respect to water pollution, the 1980 EPDA supplemented the 1974 Water Law. Water and air pollution controls both were based on national ambient quality standards, with emissions limitations imposed on an individual, source-by-source basis to meet those standards. The governor of each of Poland's 49 voivodships (administrative regions) had discretion to issue water and air emissions permits which specified emissions limitations for each individual pollution source (see Sommer, 1986).

Economic Instruments. Poland was one of the first countries to apply the 'Polluter Pays Principle' (PPP), utilizing market forces to induce reductions in pollution and resource use. And Poland's application of the PPP from the beginning was more extensive than in many western countries which employed it only for issuing permits and inspections. In Poland, 'Polluters Pay' (note the plural) always emphasized the preventative role of economic incentives, which should be more than reactive tools establishing liability for abatement or clean-up. Poland introduced its first pollution charges in the 1974 Water Law. Subsequently, under the 1980 EPDA resource use and pollution charges were applied to the whole panoply of natural resources. Per-unit charges were mandated for air pollution emissions, waste storage, water diversion and consumption, waste water discharges, use of agricultural lands for non-agricultural purposes, and for cutting down trees and bushes. These charges, as well as fines for violating environmental standards, were channeled to special funds earmarked for various purposes, including environmental protection and water management. Under a 1987 Amendment to the EPDA, one-half of collected environmental charges remained in the voivodship (administrative region), under control of the governor (*wojewoda*) for investment in regional environmental projects; the other half was transferred to Warsaw's National Fund for Environmental Protection and Water Management which invested funds in environmental projects with national significance. However, neither the National Fund nor the voivodship funds were segregated from the state budget; their operating expenses were not paid out of collected environmental charges, so that all the money collected each year could be devoted to environmental investments.

Public Participation. Polish environmental legislation appeared to be quite progressive in the 1980s, at least compared with European standards. This was particularly true of Poland's relatively extensive participation and litigation rights for ecological non-governmental organizations (NGOs). For example, unlike environmental groups in many West European countries such as Germany and Belgium, Polish environmental NGOs had the right to file public interest lawsuits in civil court (Führ and Roller, 1991).

Administrative Arrangements. In the 1960s Poland already had a central administrative agency with exclusive responsibility for air and water pollution control. In 1972 the Polish government, following the model of Britain's Department of the Environment, attempted to create an environmental 'super-department'. The Ministry for Administration, Local Management, and Environmental Protection was responsible for local administration, municipal services, and a whole array of issues related to environmental protection, including land use planning, pollution control and water management. However, this 'super-department' failed to solve the problem of coordinating governmental environmental responsibilities, and in 1983 it was replaced by a single-mission regulatory agency: the Office of Environmental Protection and Water Management. The Office was patterned to some extent on the US Environmental Protection Agency (EPA), though it also retained some economic tasks, such as water management, which conflicted with its ostensible regulatory mission. In 1987, the Office was elevated to full-fledged Cabinet status as a Ministry (with a seat on the Council of Ministers). And since 1989 it has operated as the Ministry of Environmental Protection, Natural Resources and Forestry.

Infrastructure. In Poland during the 1980s environmental awareness increased, an independent ecological movement emerged, and professional expertise developed, creating an infrastructure upon which to construct a successful program of environmental protection.

A 1983 opinion poll indicated that Poles were deeply concerned with environmental problems, which they ranked as the third most important threat facing Poland (*Polityka,* January 1984).

Poland's independent ecological movement emerged during the period of liberalization in 1980-81, and it was little affected by the period of Martial Law which lasted from the end of 1981 to 1983. During the late 1980s, Poland's ecological NGOs flourished; a number of them, most notably the renowned Polish Ecological Club, were officially registered, allowing them to take legal action for environmental protection.

In 1986 another independent environmental organization was created, the Polish Environmental Law Association. This heralded the increased importance of law and lawyers in Polish environmental protection. Two environmental law centers already had been established in the 1970s, one at Warsaw University and a second attached to the Polish Academy of Sciences in Wrocław. Each group employed academics concerned exclusively with environmental law, a specialization that was quite unusual at that time, and not only in Eastern Europe. By the mid-1980s the number of Polish lawyers dealing with environmental issues in universities, private practice and public administration was estimated to be about fifty, with another 30 or 40 working in industry as in-house lawyers, occupied mainly with voluntary environmental compensation schemes and environmental liability litigation.

All of these factors created a certain infrastructure that proved useful as Poland confronted the challenge of adapting environmental law to the needs of a democratic society and a market economy.

Overall Assessment

By the 1980s Poland's environmental problems were fairly well known and well publicized. As a result, Poland developed a reputation for being the most polluted country in the world, and it became a kind of symbol of environmental degradation (see Timberlake, 1981). Poland's disastrous ecological situation was blamed mainly on deficiencies in the institutional framework of environmental protection. Literally thousands of experts working in various environment-related fields heavily criticized government policies, accusing the government of having no comprehensive environmental policy at all. Their critiques shared two characteristics: they focused on details, and they only weakly linked legal and non-legal analyses.

Even the most ardent critics tended to take a 'hard look' only at discrete issues within their areas of expertise; they did not attempt to assess the overall regulatory program. They criticized specific shortcomings, but often considered the legislation 'in general' to be excellent. Only a few experts, mainly lawyers, exposed structural weaknesses that rendered the laws ineffective as instruments of environmental policy. These weaknesses included internal inconsistencies in the laws, the lack of judicial precision, the shortage of procedural rules, and the tendency to focus on 'ends' while ignoring 'means'. Engineers and scientists, by contrast, adjudged the laws based on their 'beauty' rather than their effectiveness because they considered the laws to be mere declarations of intent to be achieved later through appropriate administrative actions and policy measures. They were quite satisfied to have environmental laws that merely displayed some nice ideas; they were primarily concerned with influencing government policy makers. Consistent with this perspective, they tended to blame Poland's poor environmental performance on the failure of 'bad' administration to implement 'good' laws.

The revolutionary mood that seized the country when the Communists gave up power in 1989 also affected environmental protection, reversing the approach to environmental legislation: at that time, almost everyone expected radical changes in the laws. It is difficult to determine the basis of this expectation, whether it was because the entire institutional framework created by the old regime was insufficient to prevent environmental degradation or because the old environmental laws, which were designed for a centrally planned economy, would be unworkable in a market-based democracy.

The failures and deficiencies of the institutional framework were assessed not just by reference to Poland's experiences, but also in view of what other countries, such as the United States, were doing. One of the most important goals was to keep up with 'western legal standards'. A new environmental law was expected to track the latest trends in environmental policy and to employ the most sophisticated legal instruments for achieving policy goals. Among the recent trends were increased public participation in environmental decision-making and freedom of environmental information, both of which coincided with one of the basic goals of Poland's political-economic transformation: democratization. Other basic goals of the transition with application to environmental policy were 'decentralization' and 'economization'. Another, 'juridification', was particularly favored by lawyers; it conformed to the new constitutional 'rule of law' principle in Poland. A final goal more specifically related to environmental protection was 'ecologization' which, in theory, supported the introduction of 'sustainable development' throughout the economy. In reality, 'ecologization' referred only to efforts to replace the traditional 'anthropocentric' approach to environmental legislation and attempts to address 'global' environmental issues, such as ozone depletion and global warming. These goals, in one way or another, all were directed toward brand new environmental legislation. But behind them was an underlying assumption that the new law should, first and foremost, be effective and quickly improve environmental quality.

Major Challenges

The new institutional framework was supposed to resolve all of the problems that the old one either had failed to resolve (such as gaps in the system of waste regulation) or to address in the first place (such as the complete absence of environmental regulation of chemicals). The 'shopping list' for new legislation was quite long, but a couple of examples of the most hotly debated issues should suffice to give an accurate impression of experts' expectations for a new environmental law.

Enforcement. Polish environmental law was notorious for its lax enforcement; it was considered the primary problem by both foreign and domestic sources (see Jendrośka and King, 1991; Cole, 1995b). There were, however, many opinions about causes and cures for this problem. Some asked whether the old law, like many others from the communist era, was inherently unenforceable, so that it could not be effective under any circumstances.[2] This extreme view is not widely held; most scholars agree that the main problem was not enforceability, but the lack of actual enforcement (see Jendrośka, 1992; Jendrośka, 1993; Cole, 1998). And most proposals for changing Poland's environmental laws have focused on redesigning administrative arrangements to improve compliance monitoring and enforcement, particularly citizens' enforcement.

The environmental regulatory scheme under socialism vested the regional governors with primary responsibility both for issuing pollution permits and for enforcing environmental law. Since the governors' primary task, in the old system, was to satisfy the central government with their records on fulfilling socioeconomic plans, they were unlikely to jeopardize the achievement of plan targets by taking serious action to enforce environmental rules. The best cure for this seemed to be to establish a strong 'environmental watchdog', a central agency free of any economic or management tasks, responsible only for monitoring compliance with, and enforcing, environmental regulations. And legal experts recommended strong citizens' enforcement provisions to supplement, and sometimes spur, governmental enforcement.

The citizens' enforcement proposal was based on the assumption that effective environmental law enforcement depends on the number of enforcement actions, including citizens' suits. Foreign observers, in particular, noted that the number of enforcement actions in Poland was comparatively small, and they encouraged Polish lawmakers to draft provisions giving citizens the right to file public interest lawsuits (the so-called *actio popularis*) in civil court.

Out-dated Nature Conservation Law. While the 1949 Nature Protection Act had been quite progressive for the 1940s, by the 1970s it had grown out-of-date; it did not reflect the progress in scientific and legal concepts concerning nature protection. Beginning in the 1970s, numerous laws were drafted to replace the 1949 Act with more up-to-date legislation. However, none of these drafts was enacted, and the 1949 Act remained in effect, without even a single amendment, in 1989.

Lack of Effective Pollution Control Instruments. Poland's air and water pollution regimes, based on ambient quality standards, turned out to be insufficient mainly because of the weak link between the ambient standards and individual pollution permits. Polluters were not bound by the ambient standards; they only had to comply with emissions standards specified in their pollution permits. The law required polluters to apply for permits which specified emissions and effluent limitations based on ambient standards and current pollution levels. The lack of technology-based emissions controls (such as BACT or BATNEEC), as well as the broad discretionary powers of regional administrators in issuing permits, resulted in inconsistencies in setting individualized standards. This made it extremely difficult to achieve the ambient standards, especially since administrative and ecological boundaries were irrelevant to the process. This hindered the effectiveness of pollution control policy.

The proposed cure for these problems included the introduction of special water authorities and water management scheme based on catchment areas, as well as a proposal to employ technology-based approaches in granting air pollution permits.

Insufficient public involvement. Despite elaborate public participation and litigation rights, actual public involvement in environmental decision-making was quite modest. The reasons for this, besides the lack of judicial precision and the lack of specific procedural provisions in relevant laws, were related to two features of the environmental legislation: (1) lack of access to government information and (2) a preference for collective, rather than individual, participation.

In Poland, the 'environmental' community realized by the early 1980s that access to information was a prerequisite to meaningful public participation in decision-making. However, there was no general legal right of access to government information. Moreover, a presumption of secrecy was well-rooted in Poland's bureaucratic tradition. The government was always reluctant to release any specific information regarding the quality of the environment,[3] let alone information about individual polluters.

In addition to the lack of access to information, another sign of governmental aversion to accountability was its reluctance to grant individuals any meaningful participation rights in the administrative process. Broad participation rights could not be granted to the public-at-large, which was felt to be beyond the state's (and the Party's) control, but could be granted to ecological associations which supposedly were capable of control. Since, in principle, there was no room in a Communist country for independent organizations, no one expected these rights to create a 'nuisance' for the government. Indeed, even the establishment of some legally recognized, but independent NGOs did not greatly alter the effectiveness of participation and litigation rights in influencing environmental decision-making. Independent as well as officially-recognized (that is to say, Party-controlled) associations preferred the less adversarial and less demanding (in terms of legal requirements) informal means of influencing the government.[4] Experts criticized their preference for associational, rather than individual, participation rights, and they argued that the introduction of broad individual participation rights would be the best way to strengthen public participation in environmental protection. These individual rights would include not only the right to be heard (in public hearings over issuing environmental permits), but also rights to monitor compliance and enforce the law, either through public interest (*actio popularis*) civil lawsuits or as Public Guardians of the Environment, a private police force that would issue tickets (like traffic tickets) to polluters. Experts also expected that a new law would provide almost unlimited access to environmental information and documents, in accordance with 'freedom of information' and 'right-to-know' legal models.

Inefficient Economic Instruments. The rates of pollution charges and non-compliance fines in the old system were considered too low to create any incentive to reduce pollution. The way they were imposed sometimes made them counterproductive because no means were provided distinguishing between polluters who were making efforts to reduce emissions (for instance by investing in cleaner technologies) and polluters who were not undertaking any measures to improve their environmental performance. State environmental funds that subsidized pollution control did not encourage polluters to make effective use of the available funding. Subsidies also undermined what was supposed to be the basic principle of environmental policy: self-financing of pollution control.

New legislation was expected to rely on market forces to reduce pollution; economic instruments of environmental protection were expected to become as important as command-and-control measures. That required more effective pollution charges, as well as guaranteed independence for off-budget environmental funds. In addition, new legislation was expected to introduce a whole array of instruments to supplement state environmental protection subsidies, including special soft-loans, tax credits, preferential duties for pollution control devices, etc. The worldwide policy debate over emissions trading schemes, following the enactment in the United States in 1990 of a sulfur dioxide emissions trading scheme,[5] was echoed in Poland; proposals were made to employ emissions trading as part of Poland's air pollution control regime.

Developments in Environmental Law, 1989-1996

Institutional Arrangements for Law-Drafting

The Protocol of the Ecology Subgroup at the Round Table. The Round Table talks, held in the Spring of 1989 between the Communist government and the democratic opposition, laid the foundations for the transition process, including principles for a new environmental policy.[6]

The Protocol of the Ecology Subgroup included 28 goals to be realized in 1989-1990. Among them were recommendations for new economic, legal and administrative instruments of environmental protection, as well as obligations to take immediate action to close down especially dangerous facilities and to rescue certain ecologically endangered areas of the country.[7] The Protocol can be treated as an officially approved 'shopping list' for new legislation. It discussed the main deficiencies in the old framework, covered most of the issues discussed in the preceding section, and established deadlines for achieving certain policy goals. Goal number 10, for example, required the enactment of comprehensive environmental protection reform legislation by the end of 1990. To this end, a special legislative task force was to be appointed.

The Environmental Law Reform Task Force. The Task Force was appointed during the summer of 1989, and included about 70 experts on environmental issues representing academia, various governmental agencies, and ecological associations. Initially, the Task Force debated environmental policy in frequent plenary sessions, before splitting into separate working groups to deal with discrete issues. The Task Force spent much of its time debating the need for a comprehensive 'omnibus' environmental protection act. A specific question concerned whether nature conservation issues should be combined with pollution control issues in a single comprehensive statute. The Task Force prepared four working versions of new environmental legislation, each of which was subjected to an informal process of public notice and comments.[8] Out of the four working versions, a final draft was produced in 1991, but it was so watered down by compromises that hardly anyone was satisfied with it. For this reason, among others, the government decided not to wait for a new comprehensive 'omnibus' environmental protection bill; instead, it went ahead with drafting 'sectoral' laws.[9] This change in legislative policy was officially proclaimed in the White Paper on the National Environmental Policy, which was prepared by the government and approved by Parliamentary resolution in May 1991.

At the heart of the new National Environmental Policy was the concept of sustainable development. The White Paper set out the basic principles for new policies and priorities in environmental protection. And it listed a set of 'sectoral' environmental laws to be quickly enacted (see Ministry of Environmental Protection, Natural Resources and Forestry, 1991, pp. 15-16).

The Task Force wound up its efforts in 1992. Ministry officials thereafter were entrusted with redesigning the Task Force's drafts of 'omnibus' environmental and water legislation. Since then, a number of bills have been submitted to Parliament based on drafts initially prepared by the Task Force. Only two of them, however, the 1991 Nature Conservation Act and the 1991 Law on the State Environmental Protection Inspectorate, have been enacted into law. Other legislative proposals designed by the Ministry, including the draft water law, a draft waste law, and a draft law concerning environmental emergencies, have failed to gain Parliamentary approval; lawmakers found them either insufficient on the merits or incompatible with European Union Directives. These failures have led the government to search for new law-drafting arrangements. One source of funding for new legislative drafting efforts has been the European Union's PHARE program.

PHARE Projects. The EU's PHARE program, in addition to funding basic research projects concerning tradeable pollution permitting and environmental management, has funded two law-drafting projects in Poland. First, in 1994 it funded the Polish Environmental Law Association to write a draft of an environmental impact assessment law. In 1995 it funded the Polish Environmental Law Association to draft an omnibus environmental protection act.

Areas of Legislative Progress

Since 1989, Poland's existing environmental law regime has been subject to numerous changes. These changes have been introduced gradually, and in a variety of forms: by adopting new laws, amending old laws, or issuing new regulations under existing laws. The changes have not always addressed all problems sufficiently, nor have they established a comprehensive institutional framework for environmental protection. Nevertheless, the old framework seems to have been significantly redesigned with respect to nature conservation, pollution control, economic instruments for environmental protection, environmental impact assessment, and environmental law enforcement.

Nature Conservation. The long-awaited Nature Conservation Act of 1991 replaced the out-dated 1949 Nature Protection Act. The main goal of the 1991 Act is to provide a sufficient legal basis for the protection of natural resources, in accordance with the Living Nature Resources Strategy which is based on the World Conservation Strategy. To this end, the Act seeks to incorporate scientific concepts, such as biodiversity and ecosystem protection, into legal norms. Before the 1991 Act, Polish law recognized four different forms of protection: national parks, nature reserves, natural monuments, and species protection. To these the 1991 Act added several new categories, including two new large-scale forms of protected areas: landscape parks and areas of protected landscape, as well as new categories of protection for special purposes, namely, documentation sites of geological curiosities, ecologically useful sites, and landscape reserves. The Act also provides general duties for protecting wildlife, wildlife habitat, and routes of migratory animals.

The 1991 Nature Conservation Act establishes a system of nature conservation authorities, and provides them with broad powers. The general rule is that within protected areas — they now cover about 18 per cent of the total area of the country, and should ultimately cover 30 per cent — no development or economic activity may be carried out, except as authorized by the relevant nature conservation authorities, which include directors of national parks or regional governors acting through their nature conservation officers.

Despite the fact that the 1991 Act is, from a legal perspective, poorly drafted and not always sufficiently implemented, nature conservation seems to be the one area where all legislative goals have been achieved; the Polish law now meets almost all of the requirements under relevant European Union directives.

Pollution Control. Changes in Poland's pollution control scheme have been less spectacular, though nonetheless quite significant. First of all, new regulations gradually have introduced generally applicable emissions and effluent standards for air and water pollutants, respectively. In the case of air pollution, a new technology forcing mechanism establishes a timetable according to which stricter standards will apply,[10] forcing polluters, over time, to adjust their technologies to the new requirements. As a first step towards new water pollution control, a new management system based on catchment areas has been introduced. The 1990 water management regulations established seven regional Water Management Boards which are supposed to have broad powers under a new (as yet not enacted) Water Law.

Quite interesting developments have also taken place in the field of waste management. A new law, still in the drafting stage, would introduce an entirely new system of waste control, but in the meantime consecutive amendments to the 1980 EPDA have radically changed Poland's approach to transfrontier shipments of waste. In 1989 a complete ban on waste imports was introduced. That ban was lifted in 1993, and replaced with a more flexible set of regulations that implemented requirements of the international Basel Convention concerning transboundary shipments of hazardous wastes.

Economic Instruments. The 1993 Amendments to the 1980 EPDA finalized a process, begun in 1989, of separating environmental funds from the state budget. The 1993 Amendments established a three-tiered system of environmental funds: the National Fund for Environmental Protection and Water Management, voivodship funds and local funds. Both the National Fund and voivodship funds are 'arms length' legal entities reporting to independent boards; they are entitled to manage and invest their own capital, which comes mainly from their respective shares of collected pollution charges and non-compliance fines. The National Fund established its own Bank for this purpose: the Environmental Protection Bank (Bank Ochrony Środowiska) invests some of the National Fund's capital, providing low-interest loans for environmentally friendly commercial investments (see Bolan and Bochniarz, 1994, pp. 105-09). The Fund also invests some of its funds directly in private and public pollution control projects. However, the Fund's recent policy is to provide polluters with preferential credits, rather than subsidies, for environmental investments.

It is worth mentioning that Poland's rates of pollution charges and non-compliance fines have grown substantially; they are now among the highest in the world. For example, to emit one ton of sulfur dioxide now costs about $84 (USD), and to emit one ton of some toxic (carcinogenic) air pollutant, such as asbestos, costs as much as $40,000 (USD).

Environmental Impact Assessment. From 1989 to 1990 significant improvements were made in the environmental impact assessment regime that was created in the 1984 Land Use Planning Act. That law made EIAs obligatory in cases involving projects 'especially harmful for the environment and human health'. And regulations issued in 1985 listed categories of such 'especially harmful projects' (see Jendrośka and Wójcik, 1995). New regulations promulgated in 1990 added to the list of projects for which EIAs are obligatory those which 'may cause harm to the environment'. However, the regulations did not list projects that would fall within that category. The 1990 regulations also improved the EIA process by specifying detailed requirements concerning EIA content. The EIA process also was improved by the introduction, in 1989, of an EIA Commission patterned on the Dutch model (Mostert, 1994). The Commission is an advisory (only) body to the Minister of Environmental Protection, Natural Resources and Forestry; its 75 members are drawn from academia, ecological NGOs, and EIA practitioners. The role of the Commission is to evaluate EIAs prepared for 'especially harmful' projects, and assist the Minister in exercising his authority to ratify or reject the decisions of local authorities in granting planning permissions for those projects.

A new Land Use Act of 1994 further improved Poland's EIA regime by introducing a kind of 'strategic' EIA for land use plans. The Act authorizes the Environment Minister to establish two lists of categories of land use planning projects which are subject to EIA procedures; this is intended to mirror Annex 1 and Annex 2, respectively, of the EC Directive 337/85 (Jendrośka and Sommer, 1994).

Beyond all these improvements in Poland's EIA process, new draft EIA legislation is still expected to be adopted.

Enforcement. In 1991 an attempt was made to remedy Poland's environmental law enforcement deficit by strengthening the role of the State Environmental Protection Inspectorate. The Inspectorate was established in the 1980 EPDA to serve as the central agency responsible for monitoring compliance with environmental rules. The agency was, however, a 'watchdog without teeth'. The 1991 Law on the State Environmental Protection Inspectorate gives that watchdog teeth. It provides enforcement powers previously held only by regional governors; and it provides the Inspectorate with sufficient status and resources to cope with its new responsibilities.

The Inspectorate is responsible directly to the Environment Minister. It acts through a Chief Environmental Protection Inspector and regional inspectors in each voivodship. Inspectors have the power to impose non-compliance fines, to halt activities endangering the environment, and to ban the sale and import of raw materials and other products that fail to meet Polish environmental standards. The Inspectorate is also responsible for running Poland's nationwide environmental data base system. Inspectors are entitled to recover the costs of inspecting facilities (*e.g.*, sampling costs) if the inspection turns up violations.

Sustainable Development. Sustainable development was declared in 1991 to be the fundamental concept for future development in Poland. A *Sejm* resolution on ecological policy of 10 May 1991 provides that sustainable development is a prerequisite to all government programs. A similar requirement is contained in Poland's 16 December 1991 Association Agreement with the European Union. This official commitment to sustainable development has been reiterated on several occasions, including in the Governmental Implementation Program for Ecological Policy that was submitted to Parliament in 1994, and in a subsequent *Sejm* resolution of 19 January 1995 addressing sustainable development policy.

In addition to general policy statements and documents, some institutions have been established to promote, implement, and monitor sustainable development in Poland. First, the President in 1993 established an Ecological Council to serve as his advisory body on matters relating to environmental policy and sustainable development. The Council includes 50 renowned environmental experts. In 1994 the Prime Minister established a Sustainable Development Commission within the government to implement Poland's obligations under the 1992 Rio Declaration. The Commission has 30 members, representing various government agencies, who are supposed to be closely involved with implementing sustainable development policies. A third organization, the Parliamentary Working Group for Sustainable Development, was most recently established. It consists in 38 members of the *Sejm* (the lower house of Parliament) and 6 members of the Senate (the Parliament's upper house). The group's role is to monitor and coordinate activities of various parliamentary committees and subcommittees from the point of view of implementing Agenda 21 which, like the Rio Declaration, was signed by Poland at the 1992 Earth Summit in Rio de Janeiro, Brazil.

However, the rcal importance of the concept of sustainable development can be seen from the fact that it originally was *not* introduced in the 'Strategy for Poland', which laid down a set of priorities concerning the future direction of national socioeconomic development. To that end, it established 10 programs to achieve goals in fields considered to be of crucial importance for the socioeconomic development of Poland. Environmental protection and sustainable development were not among the 10 programs. But this oversight was rectified two years later when the 'Strategy for Poland' was supplemented with an 11th program relating to environmental protection and sustainable development. However, program number 11 has been heavily criticized for being nothing more than a list of measures (mostly investments) to protect the environment; it provides no guidelines for sustainable development.

Generally speaking, Poland's efforts to promote and implement sustainable development seem quite significant. This is especially evident in the 'Green Lungs of Europe' program, which Poland initiated. In February 1993 seven countries signed on to the program, which commits a huge unspoilt region of Eastern Europe to sustainable development. This marks Poland's strongest commitment to implementing sustainable development, a commitment that has been appreciated by western observers (see Cole, 1995b, pp. 313-14).

Perspectives

New Laws in the Pipeline

The Polish Parliament recently considered a private bill on pollution prevention that would introduce product charges, a toxic release inventory, and numerous economic incentives to promote clean production (over traditional end-of-the-pipe technological controls). In addition, the government is just about to approve a draft Environmental Impact Assessment (EIA) law that would establish 'strategic' EIAs for government plans, programs and legislative proposals. It would also provide for fully-fledged EIA procedures for concrete development projects, stressing public participation which currently is provided only at the screening stage. The draft EIA law is meant to bring Poland's EIA scheme into full compliance with the EU's EIA Directive 327/85 (bearing in mind pending amendments to that Directive) and the Espoo Convention.

With respect to the draft 'omnibus' environmental protection law, the government is requiring that new Polish environmental legislation must comply with all EU laws, especially those relating to access to information, waste management, and air pollution control. In particular, new 'omnibus' environmental legislation is expected to facilitate the future introduction of integrated pollution control in Poland. The draft law also will attempt to institute procedures for negotiated permitting.

Designing a New Environmental Policy 'Tool Kit'

Command-and-control Instruments. The traditional command-and-control approach remains at the core of Poland's environmental protection program, and even the most far-reaching and progressive proposals would not change that focus; other categories of instruments will serve only as supplements. There are, however, on-going discussions about replacing certain command-and-control instruments with more efficient mechanisms for environmental protection.

The existing pollution control program is based on ambient environmental quality standards and emissions limitations set individually for sources in separate permits for water pollution, air pollution, waste disposal, etc. This program has been heavily criticized for its media-specific approach, which is both inefficient economically and ineffective environmentally, and for its individualized emissions limitations, which in practice turn out to be difficult to administer and enforce by agencies that are understaffed, underfinanced, and undertrained.[11] To this point, regulatory changes have been limited to introducing additional instruments and enhancing existing ones; heavily criticized aspects of the law remain unchanged. In fact, there is hardly any concept for a tentative new regulatory program of pollution control. So far, only some very general discussions have taken place, and a few initial steps have been taken to replace individualized emissions limitations with generally applicable limits.[12] There has been no progress at all on instituting integrated pollution control, despite earlier overly optimistic expectations (see Jendrośka, 1993, p. 108). Moreover, the Implementation Scheme for the National Environmental Policy (adopted by parliamentary resolution on 19 January 1995) does not even mention integrated pollution permitting. In fact, integrated pollution control and other regulatory reforms are impeded by a fundamental political obstacle: meaningful reforms would affect entrenched interests within the environmental agencies, requiring administrators to accept new duties and to develop new skills. As with any administrative reform that infringes on the interests of administrators and their 'constituencies', efforts to reform environmental policy tend to get stuck in a bureaucratic maze.[13]

Voluntary Agreements. In recent years, the attention of environmental policy makers all over the world has been drawn to voluntary agreements. These agreements are founded on consensus-building and cooperation between industry, government and the public. The Netherlands was the first country to experiment with voluntary agreements for pollution control, and they heralded a profound change in that country's administrative culture (van Zijst, 1993, p. 13). The Dutch have utilized two forms of voluntary agreements: covenants, which are agreements between industry and the government; and contracts, which are agreements between polluters and affected private parties (van Dunne, 1993). The success of these voluntary agreements has made the Dutch model very popular with policy-makers in other countries, some of which have already introduced similar instruments. Even in the United States, a country notorious for its adversarial approach to environmental protection, so-called 'Good Neighbor Agreements' between companies and local communities are becoming increasingly popular (Levis, 1994, pp. 143-61).

Poland does not seem to be in a position to follow the Dutch example, even though its environmental protection regime suffers from precisely the same administrative and enforcement problems that led Dutch authorities to introduce voluntary agreements in the first place (see Glasbergen, 1994). There are several reasons for this. Voluntary agreements must be supported by some level of enforcement. In Poland, they would have to be enforced in civil courts, which are the weakest branch of its judiciary. Moreover, voluntary agreements require a certain level of stability among the parties. During Poland's ongoing transition to market democracy, however, ownership changes and industrial restructuring do not provide much stability. Industrial organizations are not sufficient well-established to serve as responsible and reliable partners in voluntary agreements with the government.

This does not mean, however, that Poland cannot or will not embrace a consensual approach to environmental policy. In fact, Poland's commitment to that approach is reflected in the government's proposal for 'negotiated environmental permitting', which is intended to supplement, though not replace, the 'usual' permitting procedures. The proposal envisions allowing applicants, in certain kinds of cases, to request negotiations with the government about permit conditions, including individualized emissions limitations, time schedules for compliance, rates of pollution charges, etc. The proposal would establish special 'negotiating commissions' composed of representatives from the environmental permitting authority, environmental law enforcement agencies, and local authorities. The results of negotiations between the negotiating commission and permittees would be binding on both parties.

There is, however, a principal difference between this type of negotiated permitting and the Dutch voluntary agreements (covenants and contracts): negotiated permitting invokes rules of administrative procedure, rather than rules of civil procedure. In this respect, the concept of negotiated permitting more closely resembles the American model of negotiated rule-making, than the Dutch model of voluntary agreements. This does not make the proposal any more likely to be adopted, however, since negotiated permitting requires relations between administrative agencies and the regulated community that are still foreign to Poland's administrative culture.[14]

Instruments Based on the Public Accountability of Polluters. Since the 1980s, many western countries have shown a tendency to deregulate. This tendency is reflected, in part, in the increasing emphasis placed on promoting and fostering eco-management within industries. Various schemes and instruments have been invented to strengthen environmental ethics, and to encourage voluntary efforts to reduce pollution. All these schemes and instruments rely heavily on the public accountability of polluters who are driven by market forces. Despite some concern about the applicability of market-based eco-management techniques to Central and Eastern European economies which only recently have become 'marketized' (Greenspan Bell, 1991), Poles are beginning to realize the 'good' as well as the 'bad' traits of consumerism, including their power as consumers to drive the market to provide 'environmentally friendly' products.[15] However, in the absence of any regulatory scheme to certify such products, producers can, and are, taking advantage of consumers by misrepresenting products as 'environmentally friendly'.

Polish policy-makers have not yet had time to examine the potential role that the public accountability of polluters can play in achieving environmental goals; no policy measures have been introduced to encourage eco-management among industries. The Polish government's lack of attention to the utility of public accountability for environmental protection is reflected in its attitude toward a so-called 'List of Major Polluters', which was supposed to be made public so that the polluters would be subject 'to intensive supervision by the environmental authorities *and the public*' (National Environmental Policy of Poland, 1991, p. 12, emphasis added). This scheme might have had effects similar (in function, if not in scope) of the Toxics Release Inventory in the United States which garnered so many voluntary pollution reductions that some commentators consider it to have had 'more impact than regulatory programs' (*The Nation's Polluters,* 1991). Unfortunately, Poland's 'List of Major Polluters' was placed under the 'supervision by the environmental authorities', who refused to make the list public, thereby thwarting the potential environmental improvements that might have resulted from public accountability.[16]

Only recently has the Polish government unveiled a proposal to introduce 'eco-labeling' for products. This proposal is supposed to conform to European Union legislation. However, Poland's proposal lacks any meaningful provisions concerning verification of labeled products, and there are no corresponding proposals for 'eco-auditing'. For these reasons, the current proposal seems to be mere lip service, rather than a serious effort to introduce public accountability for environmental protection. Poland's environmental policy makers still do not seem to believe in industry self-regulation (or, more accurately, information-based market regulation of polluters). Their caution is documented in the current Implementation Scheme for the National Environmental Policy which touches on the issue of public accountability only in describing Poland's pending 'Clean Technology' program.

Economic Instruments. Given the Polish government's attitude toward other environmental protection instruments, it is not surprising that Poland's policy-makers have focused their attention over the past couple of years on economic instruments for environmental protection, specifically ecological charges (fees and fines). There are many reasons why Polish policy-makers prefer such economic instruments: Poland has a long history of using ecological charges, and a well-established institutional framework for assessing and collecting charges; economic instruments are relatively easy to administer; they have proven (in recent years) to be effective as pollution control devices; and they are economically efficient. Perhaps more important than all these reasons, however, is the fact that collected ecological charges are a huge source of off-budget revenue for environmental funds, benefiting the political and administrative forces associated with those funds. Notwithstanding the fact that Poland's environmental funds are administered by arm's-length boards of directors, they remain in the domain of the Environment Ministry, and they enhance the prestige and power of the Environment Ministry in the eyes of politicians and administrators.

The Alchemy of Law-drafting

Neither Quick Nor Easy

It was clear already in 1991 that hopes of completely restructuring Poland's environmental institutional framework within two or three short years — let alone in the time frame envisioned during the 1989 Round Table negotiations — had been overly optimistic.

The idea of radical change was abandoned not so much because of a lack of political will or because environmental protection was not a priority during the transition period, but rather because radical change proved to be impracticable. For one thing, legislators quickly discovered that the environmental protection regime is not self-contained, but is closely interconnected with the entire legal system. For another, Poland's available human resources (which were, in fact, quite sizeable by 'Eastern' standards) were insufficient to revise the law thoroughly within such a short period of time. Their efforts were laudable, but law drafters encountered many dilemmas which required various legislative and institutional solutions.

Problems Encountered

Coordination. The transition period, with its historic aim of establishing a completely new economic and political system, was considered by many to provide an excellent opportunity for introducing highly progressive environmental policies and laws. Indeed, the transition period provided a certain bias toward progressive thinking and radical change, which gave policy-makers and law drafters much more room for creativity than more stable political situations tend to afford. By nature, more established and stable systems, democratic or otherwise, tend to be more conservative; they are less likely to produce radical reforms. On the other hand, stable political-economic environments at least offer a firm basis for reforms, which is a huge advantage because new schemes cannot not be developed in a vacuum. Even the most radical policy changes, in order to be operational, must be facilitated by an overarching institutional framework. From this perspective, contrary to expectations, the transition period definitely has not been a paradise for policy-makers and law drafters. In fact, as a purely technical matter, policy-making and law drafting are a nightmare in circumstances in which policies are being reoriented and laws are being redesigned all at the same time.

The collapse of the Communist regime caught Poland's democratic opposition by surprise. Unlike opposition groups in democracies, which are accustomed to moving in and out of power, the Polish opposition had no previous experience of power; it was not even in a position to prepare any general plans for governing, let alone comprehensive and detailed programs for change, before it actually took over power in 1989. It was unreasonable to expect otherwise, given the opposition's lack of access to information, its lack of government experience and expertise, its internal political diversity, and the huge tasks confronting it.

New laws and policies relating to all matters, ranging from the constitution to local administration, were being worked out all at once by myriad legislative committees and task forces. Since various laws within a single piece of legislation had to be, at least, compatible, there had to be a great deal of coordination on interdependent issues. Coordination, however, turned out to be the weakest link in the legislative process.

Perhaps the best example of coordination problems in environmental legislation concerned the need to adjust the powers and structures of environmental administration in light of changes made to the overall framework of administration in Poland. These changes, especially the process of decentralization, have ebbed and flowed; many are far from complete. They include the restructuring of central government, the reorganization (that is, enlargement) of regional governmental administration, and the establishment of a second tier of local self-government. The lack of progress on these administrative changes still hinders serious discussion about related changes in environmental administration.

Insufficient knowledge. Ambitious goals were imposed on law drafters without adequate consideration of whether the existing environmental infrastructure could support them; there were real questions, for example, about whether Polish environmental law experts possessed the background knowledge and expertise needed to design and implement instruments to achieve the goals. In fact, much of the available domestic expertise turned out to be irrelevant for the task mainly because Polish experts had little familiarity with foreign systems of environmental protection. Few Polish environmental lawyers, for example, had established working relationships with Western environmental law experts;

most had little more than a rough idea of 'Western' laws and ecological concepts. And yet they were expected to provide guidance for environmental policy-makers and law drafters in Poland.

Much of the initial impetus for drafting new environmental laws came from independent experts who tried, without success, to persuade the government to implement their ideas during the 1980s. But their old 'shelved' drafts proved to be of little use after 1989, when the needs for new environmental legislation were vast and time was short. The old draft laws covered a few important issues but with respect to many other outstanding issues, they provided little guidance; there was hardly any background information to facilitate the drafting of complex laws and regulations in a short period of time. That background information had to be developed before the law drafting could even begin; basic research needed to be conducted, and studies needed to be written. The scope of the task varied with the specific issue. For example, with respect to basic pollution control issues, Poland's domestic situation already had been thoroughly assessed, but there were no studies comparing solutions applied in other countries. Those studies had to be carried out so that Polish law drafters would have some basis for comparing potential solutions for shortcomings in Poland's environmental laws. With respect to other issues, such as the environmental regulation of chemicals, which had only recently been identified as a problem in Poland, there was little available information about the domestic situation, let alone the situation in other countries. Finally, there were issues, such as access to environmental information, where Polish experts already had studied both the domestic situation and foreign solutions, but the legislation could not be drafted and implemented without corresponding changes in other laws.

New Issues. The economic and political transition itself presented several additional problems for law drafters, including deregulation, privatization, and marketization.

Deregulation. The most hotly debated environmental problem caused by deregulation arose shortly after the 1988 Law on Economic Activity took effect in January 1990.[17] In addition to freeing virtually the entire domestic economy from central control, that law removed barriers to foreign trade. Prior to 1988 only a few state-owned enterprises were authorized to carry on foreign trade, and they were subject to stringent limitations; their licenses specified the categories of goods or materials they were permitted to trade. Waste products were not listed in any license, so wastes could not legally be shipped to or from Poland. However, the 1988 Law on Economic Activity liberalized foreign trade, abolishing foreign trade licensing. No one anticipated the (unintentional) massive influx of waste imports into Poland that resulted, as small companies, established exclusively for the purpose of importing waste, brought tons of sometimes hazardous waste into Poland and, in most cases, dumped it illegally. This presented a new challenge for Polish environmental law: in the absence of general restrictions on foreign trade in Poland's new economic system, Polish experts had to design a new regulatory program specifically to control the transfrontier shipment of waste.

Privatization. Privatization created new and special problems concerning environmental liability in property transfers. Lawmakers did not address these problems in the privatization law or draft environmental protection legislation. And the Polish government was surprised when prospective investors began raising environmental questions (see Greenspan Bell, 1992, p. 10095). Two *ad hoc* solutions were devised to deal with these questions: (1) the Privatization and Environment Ministries created an interministerial task force to deal with environmental issues in privatization, and (2) the Privatization Ministry began employing expert Western advisors to negotiate environmental liability issues in privatization deals with foreign investors. Each of these solutions has proven to be effective. They undoubtably will contribute to the development of an overall legal scheme for dealing with contaminated land, which so far has been a somewhat neglected issue in environmental legislation (see Wajda and Sommer, 1994, p. 194).

Marketization. Like privatization, the marketization of the Polish economy has created new environmental issues. Poland's free market economic system is consumer-driven, and consumers have been driving a shift away from production of investment goods to production of consumer goods. This shift has resulted in significant increases in waste production, and changes in the nature of waste being produced — there is less industrial waste but more household waste. This requires a change in waste management practices because Poland's old waste control system was geared toward industrial waste problems.

Legal Tradition. Public participation seems to be the only area of environmental protection in which absolutely no progress has been made, despite the fact that a comprehensive draft law incorporates the most advanced Western provisions for public participation in environmental decision-making. The reasons for this go beyond the natural bias of bureaucrats against public interference in their business.

Public participation and access to government information are necessary elements for any contemporary environmental regulatory regime. But they reflect a conception of 'open' government which is foreign to the Polish legal culture. Poland's administrative system is based on German, Austrian and Russian traditions (stemming from the 123 year period when Polish territories were partitioned among those three powers). Each of those traditions is notorious for its secrecy and aversion to public involvement. Poland's approach to administrative procedure, in particular, is based on a nineteenth-century German model in which the state administrative authority represents the public interest, while parties participating in administrative proceedings represent their own private (individual or group) interests. Consequently, for rule-making this model prefers 'cabinet' proceedings where only experts are invited to participate. Because Poland's administrative system is constructed on this model, it is not easy to introduce general public participation rights for environmental protection. The concept of public hearings, in particular, does not fit easily into Poland's administrative model; the public hearing is an invention of a very different administrative system within which participation is viewed as an individual right, proceedings are adversarial, and the administrative authority presides as a judge. But these conceptual distinctions are not the main reasons why Polish authorities have been so reluctant to provide public hearings for environmental issues. There are two more important reasons: first, Polish administrators are disdainful of the views of laypersons; second, they have a strong aversion to undertaking additional duties. This attitude unfortunately is reinforced by a common misunderstanding of the role of public participation in environmental decision-making. Government officials tend to view public participation in the extreme, as a situation in which citizens' comments are binding on environmental decision-makers (see Jendrośka and Jerzmański, 1995).

With respect to access to environmental information, there is another problem: the lack of accompanying legislation without which any meaningful 'freedom of information' (FOI) legislation is impracticable. This includes, in particular, trade secret legislation and legislation protecting personal data, which either are completely absent in Poland or insufficient for Poland's new market democracy.

Aside from these general problems, there are certain other concepts which have yet to be introduced into Polish environmental law. They include technology-based approaches, such as 'best available technology' (BAT) and 'best environmental practice' (BEP). The introduction of both is vitally important in order to harmonize Polish environmental laws with EU legislation, as well as international law obligations such as the new Helsinki Convention. BEP, in particular, is a concept that will require a certain shift from current practice in Poland; it involves energy and resource conservation measures, recycling, clean technology, and product controls. All of these instruments are relatively new for the Polish framework, though law drafters have long planned to introduce them. Nevertheless, it is certain that their introduction will require a complete reformation of Polish environmental law.

Organization and skills of the civil service. Poland lacks trained, organized, and adequately equipped civil service personnel. Of course, this problem extends well beyond environmental administration, and was caused by the political status of bureaucrats under communism. So far, during the transition period politicians have been unable to approve clear legal rules to ensure the apolitical status of the civil service, or the clear division of powers and duties among agencies. As a consequence, efforts to transform lazy communist bureaucrats into efficient civil servants has not been successfully completed. An associated problem is the highly politicized structure of the administration, which creates a 'winner takes all' rule: the party (or parties) that prevail in government elections take over the entire bureaucracy. In this context, it should be noted that there have been six different governments and five different Environment Ministers (so far) in Poland since 1989. Such frequent cabinet changes certainly do not contribute to the stability of Poland's environmental policies. And there is no mechanism to assure administrative continuity. Newly appointed administrators usually demand to review each issue and policy *ab ovo*. This has profound consequences for the way in which environmental law is understood, designed, implemented and enforced in Poland.

Legislative Policy Dilemmas

A number of dilemmas have arisen during the process of drafting new environmental laws and amendments, all demanding some kind of solution. At least some of them are worth mentioning (in no particular order of importance).

Judicial precision vs. common understanding. There is a long tradition in Poland, dating back at least to the 16th century, of strong individual rights and a disdain for law.[18] The traditional disdain for law (though not the tradition of strong individual rights) was reinforced by communist ideology which subordinated legal rules to political and economic expediency. In the case of environmental law, this general attitude coincided with a tradition, believed to be common throughout Europe (see Smith, 1992, p. 40), of treating environmental matters as merely 'technical' rather than 'legal' problems. In Poland, for example, environmental laws traditionally have been drafted, implemented and even enforced by non-lawyers.[19] Consequently, neither industry nor government has felt bound by the laws; rather, they have treated the laws as mere 'guidances'.

The legislative dilemma was whether to draft the laws in precise legal language and risk having it misread or neglected by agencies composed of non-lawyers, or continue the abhorrent tradition of drafting imprecise, incoherent laws with at least some hope of having the main policy goals implemented.

A strong pressure to enact short, laconic and simple laws conflicted with a desire to regulate precisely all details within the statute, delegating as little discretion as possible to distrusted administrators. That distrust was a vestige of Communist times, and the 'us versus them' mentality that prevailed between the citizens and the Party/state. Law drafters' desires to reduce administrative discretion also conflicted with a perceived need to provide more flexibility in environmental regulations. For example, lawmakers wanted to treat polluters who undertook voluntary environmental protection initiatives differently from more recalcitrant polluters who did nothing at all, but that would have required giving more discretionary authority to administrators.

Enforceability versus progressive policy. Many dilemmas arose from the contradictory character of certain goals set forth in new legislative proposals. Old legislation was blamed for setting lax standards and for general unenforceability. New legislation was expected to be both more ecologically progressive, stringent, *and* more fully enforceable. But achieving all those goals, under the circumstances, seemed impossible. Consequently, law drafters opted for enforceability over stringency.

Ideal solutions versus reality. One of the biggest dilemmas that confronted lawmakers was whether to follow the tradition of drafting nice-looking, comprehensive, modern laws that could proudly be presented to the world, or to focus on drafting less 'fancy' but workable laws. The latter approach required more attention to be paid to administration (and its limits), the economic situation of polluters, and the compatibility of environmental laws with the rest of Poland's institutional framework. This more 'pragmatic' approach also presumed that new legislation would make use of pre-existing institutions and procedures, rather than constructing new ones.

Decentralization versus effectiveness. The trend toward decentralization in Polish administration was countered by the need for well-qualified professionals to administer environmental programs; such professionals were unlikely to be found in small self-governing communities.

Environmental code versus sectoral laws. Perhaps the most hotly debated, though not the most important, topic concerned whether to continue tradition by attempting to draft another 'omnibus' code-like environmental law, or to proceed with enacting 'sectoral' laws as soon as they were ready. The biggest disadvantage of the first approach was that it required consensus on many issues at once; a single controversial issue could undermine the entire project.

Approaches Employed

Law drafters in Poland took a very cautious approach to new instruments and institutions of environmental protection. They weighed needs against possibilities in order to avoid establishing new requirements that industry could not meet or the government could not administer. Concern over potential administrative problems, in particular, derailed badly needed legislation to regulate chemical manufacture, use and disposal. Other proposals that in the current situation might prove either unenforceable, such as more stringent emissions and effluent standards,[20] or counterproductive, such as the immediate introduction of technology-based (BAT or BATNEEC) goals, were postponed.[21] However, the Polish government gradually has been introducing technology-based emissions standards.

The process of environmental law reform also has been slowed by the need to build a professional infrastructure through the gradual implementation of programs, such as the EIA regime, or pilot projects to experiment with certain types of measures, such as emissions trading (see Dudek et al., 1992).

Mistakes

Many of the errors made in designing a new legal framework for environmental protection originated from mistaken assessments of the existing framework. First of all, Poland's existing environmental laws potentially were much more effective than many believed; the real problem was Poland's inefficient economy within which those laws operated (or, more accurately, could not operate). Mistaken assessments of existing environmental laws were based on two false assumptions. First, Poland's environmental laws were compared unfairly and unfavorably with modern Western environmental laws — a fairer, more realistic, assessment, based on factors such as technological capacities, would have compared Polish laws with western laws from the 1950s and '60s. Second, Poland's environmental protection regime was assessed only on the basis of its 1980 EPDA, while other instruments provided many effective control functions, for example concerning transfrontier waste shipments.

Polish legal experts spent a great deal of time trying to cure 'old' problems, but overlooked new ones. For example, the State Environmental Protection Inspectorate was given broad power to enforce environmental rules because regional governors, who in Poland's former state-owned economy were primarily responsible for regional production levels, understandably failed to enforce environmental rules against enterprises in their regions. But privatization broke the links between ownership (and control) of the means of production and environmental regulators, and there is no other important reason to separate pollution permitting from such administrative functions as compliance monitoring and enforcement. It is now necessary to distinguish activities that should be subject to *local* licensing, monitoring, and enforcement from those that should be subject to *central* control.

The legal experts were preoccupied searching for legislative solutions to problems that could be (and were) solved by other, non-legislative means. For example, a great deal of effort was expended to strengthen public participation in environmental decision-making by establishing individual rights; meanwhile, increased public participation followed as a practical consequence of two unrelated developments: Poland's old restrictive association law was broadened, and Polish NGOs became more active in international fundraising. Because of these developments, there are more associations in Poland that can participate in environmental decision-making, and they have a stronger financial footing for participation. Poland's system of public participation now functions similarly to the system in the United States: in both countries, it is mainly environmental NGOs that take advantage of existing participation rights, even though in the US (but not in Poland) individuals as well as associations have *legal* rights to participate.

Polish experts assumed that privatization automatically would improve compliance with environmental regulations; enforcement was expected to be made easier once environmental regulators no longer had direct economic ties to the regulated industries (as they did under socialism). In fact, however, the reverse has been the case. State-owned firms under socialism were not in a position to neglect the law, at least not overtly, because they were effectively controlled by the Communist Party. Today, there is nothing like 'Communist Party control' to supplement inefficient governmental monitoring and enforcement; private companies in many cases do not bother with environmental requirements. Polish lawyers had hoped that enforcement in post-Communist Poland could be effectively supplemented with citizens' suits in civil courts. But that hope was misplaced because it failed to take into account at least two significant differences between the US judicial system (where citizen suits were pioneered) and Poland's judicial system. First of all, while in the United States judicial review of administrative action is undertaken in the civil courts, in Poland administrative oversight is channeled through the administrative courts, which are a separate and independent branch of the judiciary. Moreover, unlike in the United States, there is no *pro bono* practice, and (as in England) there is a rule that the losing party bears all the litigation costs and attorneys' fees. Thus, for citizens to seek administrative redress in civil court in Poland is a comparatively risky and potentially expensive option.[22]

In Search of Cost-Effectiveness. A mature market economy, which Poland claims to have achieved, requires cost-effective public policies, including environmental policies. While Poland's environmental policies have triggered a substantial increase in environmental investment expenditures in recent years, making those investments cost-effective still presents a challenge (see Żylicz, 1994, p. 79). Neither the overall institutional framework nor specific environmental policy instruments have been assessed for their economic efficiency. In fact, Poland's transition to market democracy was already in its fourth or fifth year before environmental policy makers started thinking in 'market' terms. The government just recently commissioned studies to determine the cost-effectiveness of current environmental policies. This reflects a more general change in attitude towards environmental policy in Poland; environmental effectiveness is no longer sufficient, policy instruments also must be economically efficient.

Following the experiences of the most environmentally advanced western countries, Polish policy-makers today face the challenge of designing a new environmental policy 'tool kit' that can achieve ambitious policy goals, providing decent environmental conditions, but at a reasonable price.

Conclusions

Does It Work?

There is no way to assess whether more radical reforms would have improved Poland's environmental performance more, or more quickly, than the cautious reforms actually undertaken. It is possible, however, to monitor whether Poland's reformed policies and legislation are being successfully implemented in order to test claims made by some critics (such as Manser, 1993) who blame Poland for failing to introduce radical changes in environmental policy, for missing a chance to reorient economic policies for environmental protection, and for having made no progress in environmental performance.

The available evidence refutes those accusations. Pollution levels in Poland have dropped since 1989. Those reductions to a large extent are attributable to closing down old, out-dated, energy consuming and highly polluting heavy industrial facilities. But Poland's new environmental policy also has contributed significantly to this progress. It must be borne in mind that Poland only recently attained a level of spending on environmental protection commensurate with what is spent in the OECD countries. Poland's current environmental investment rate of about 1.3 per cent of GNP corresponds, in absolute dollar terms, to the level of per capita environmental spending in western countries at the time they began their own large-scale environmental recovery programs in the 1960s and 1970s (Żylicz, 1994, pp. 80-1).

While Poland's environmental statistics show certain improvement, they are not worth much when it comes to assessing the effectiveness of Poland's institutional framework for environmental protection. That is better assessed by applying the criteria of flexibility and amenability to modern environmental policy instruments. On these criteria, Poland's legislative changes appear to be moving in the right direction, only slowly.

Lessons Learned

Poland has not radically reformed its environmental laws, if 'radical reform' is defined as adopting a wholly new regulatory scheme and administrative framework. Moreover, the experience of the past five years seems to show that radical reform was neither feasible nor advisable. It is a cliché to assert that legislative reform requires knowledge and time, and Polish lawmakers did not possess nearly enough of either. So initial hopes that Poland might 'leapfrog' the environmental policies of the West were overly optimistic. Comprehensive, modern environmental protection systems require a certain infrastructure based on knowledge, technology, administration, and economic structures. These factors can only be developed hand-in-hand.

Environmental legislation, in order to be effective, must fit within a legal tradition and an administrative framework. Therefore, as Poland's experience shows, foreign assistance is most effective when focused not on drafting new laws, but on providing law drafters with information about specific issues and how those issues are handled in advanced systems of environmental protection. Foreign experience and expertise are crucial, but not just so that Poland might mimic foreign laws; to be useful those laws must be adaptable to Polish needs, legal traditions and institutions. This requires an understanding of the foreign systems and institutions within which those environmental laws operate. It is not enough that a foreign environmental law works well; Polish law drafters must be able to identify why it works well in the other country's system, and then determine whether it is compatible with, or adaptable to, the Polish system.

Poland was fortunate to have a comparatively well-developed regime of environmental protection in place before 1989. The failure of that regime under Communist rule has been attributed, somewhat naively, to lax enforcement (see Jendrośka and King, 1991). The main reason, however, was more fundamental: the overall inefficiency of the Communist regime (see Sommer, 1993, p. 25).

Poland's regulatory framework has changed gradually as certain questions have been answered: Was the old regime of environmental protection designed to fit only in the old political-economic system?[23] Which instruments worked in the old system, and could they also work in the new one? And which instruments did not work in the old system, and would they work better in the new one?[24]

After five years of gradual changes, Poland's environmental protection regime is significantly restructured. New background knowledge has been gained; new infrastructure has been developed; and new procedures have been introduced to assist in adjusting Poland's environmental laws to meet European Union requirements.[25]

Notes

1. In Germany, the 'environmental code' has been officially adopted as a matter of policy, and the process of drafting such a code has begun. However, in other countries, the concept of an environmental code is still a matter of theoretical discussion.
2. The question of the enforceability of Polish environmental law is broadly discussed in Cole (1998). Cole concludes that environmental laws enacted in People's Poland were *not* unenforceable.
3. The famous case where the government failed to inform the public about the Chernobyl nuclear accident is just the most familiar example of the common practice of withholding information.
4. For a more detailed analysis of the reasons why ecological organizations and associations did not exercise public participation rights before 1989, see Jendrośka (1991a, pp. 231-7).
5. 42 U.S.C. § 7651.
6. It is worth noting that the Ecology Subgroup talks arguably were the least adversarial of all the 14 subgroups organized to negotiate various aspects of the democratic transition. Except for divergent views about the future of nuclear energy in Poland, there were no drastic differences between the two sides with respect to ecology and natural resources management. Both ecological teams were composed of academics and experts committed to environmental protection; their positioning on opposite sides of the table was artificial.
7. An English translation of the Protocol of the Ecology Subgroup, and a commentary, can be found in Bolan and Bochniarz (1994).
8. Notice and comment procedures under Poland's 1987 Referendum and Consultation Act are not comparable with more formal rule-making procedures provided in more sophisticated laws, such as the US Administrative Procedures Act, 5 U.S.C. § 551 *et seq.*
9. For more information on the draft omnibus environmental law and the draft water law, see Bowman and Hunter (1992).
10. The 1990 air pollution regulations set forth emissions limitations that apply as of January 1994, more stringent standards apply as of 1997, and so on.

11. From the point of view of resource allocation, however, individualized emissions limitations are considered economically efficient.
12. Categorical emissions limitations have been imposed on internal combustion engines. With respect to stationary sources, categorical limitations have been imposed on the energy industry for emissions of sulfur dioxide, nitrogen oxides and dust. Under the Water Law, generally applicable discharge limits are being introduced gradually.
13. In this respect, Poland's situation resembles the well-researched problem in the US, where environmental agencies and their constituents, the regulated communities, are reluctant to upset the stable relationship they have development (see Buck, 1991, p. 44).
14. This is reflected in the fact that the State Environmental Protection Inspectorate, Poland's chief environmental law enforcement agency, has shown much more support for negotiated environmental permitting than the state agencies that actually are responsible for permitting.
15. A series of opinion polls commissioned in 1994 by various ecological and consumer groups showed that 70 to 80 per cent of Polish consumers were will to pay up to 10 per cent more for environmentally friendly products, while 20 per cent were willing to pay even significantly more than that.
16. For detailed data concerning the environmental performance of Poland's 'Major Polluters', see Bolan and Bochniarz (1994, pp. 237-94).
17. It is worth noting that the 1988 Law on Economic Activity was the Communist regime's final and most radical step to reform the socialist economy. It deregulated virtually the entire economy, introducing almost total freedom among economic actors. Since then, public control has gradually been reintroduced mainly for environmental and public health reasons.
18. The disregard for law may be a consequence of lax enforcement, which has been a problem throughout Poland's history. Indeed, during the eighteenth century, private enforcement was a semi-officially accepted practice.
19. It is not at all unusual in Poland for a non-lawyer to sit as chief of an appellate review board within a ministry, even though such boards deal exclusively with legal issues.
20. Poland's decision to ratchet up its emissions standards in stages over a period of years implicitly recognizes the need for enforceability, as well as the need for more stringent environmental standards.
21. A survey of central and regional level environmental administrators indicated that they are not yet prepared to implement a technology-based approach, and that the forced implementation of BAT or BATNEEC requirements would introduce so much chaos into the permitting process that the measures would prove counterproductive.
22. It is worth noting that these rules do *not* apply in Poland's administrative courts, so citizens who sue to overturn administrative rulings in administrative courts do not face the same risk of loss if they lose a case.
23. Some maintain, in fact, that the 1980 EPDA is better suited to the new market-based system than it ever was to socialism (Cole, 1995a, p. 63).

24. For example, economic incentives could not work in a non-market economy, but they seem likely to work in the new system (see generally Cole, 1995b).
25. For example, a detailed procedure recently was introduced to check, in a routine way, how legislative proposals and draft laws accord with European Union requirements.

References

Adams, T.L. and M.E. Cox (1990), 'The Environmental Shell Game and the Need for Codification', *Environmental Law Reporter, News and Analysis*, vol. 20.

Bell, R. Greenspan (1993), 'Exporting Environmental Protection', *Environmental Law Reporter*, vol. 23, pp. 10701-10704.

Bell, R. Greenspan (1992), 'Industrial Privatization and the Environment in Poland', *Environmental Law Reporter*, vol. 22.

Bolan, R.S. and Z. Bochniarz (eds) (1994), *Poland's Path to Sustainable Development 1989-1993*, Hubert H. Humphrey Institute of Public Affairs, Minneapolis.

Bowman, M. and D. Hunter (1992), 'Environmental Reforms in Post-Communist Central Europe: From High Hopes to Hard Reality', *Michigan Journal of International Law*, vol. 13, pp. 921-980.

Buck, S.J. (1991), *Understanding Environmental Administration and Law*, Island Press, Washington, D.C.

Cole, D. (1995a), 'An Outline History of Environmental Law and Administration in Poland', *Hastings International and Comparative Law Review*, vol. 18, pp. 297-358.

Cole, D. (1995b), 'Poland's Progress: Environmental Protection in a Period of Transition', *The Parker School Journal of East European Law*, vol. 2, pp. 279-319.

Cole, D. (1998), *Instituting Environmental Protection: From Red to Green in Poland*, Macmillan, Basingstoke, UK.

Dudek, D., J. Kulczyński and T. Żylicz (1992), 'Implementing tradeable rights in Poland: A case study of Chorzów', paper presented at the Third Annual Conference of the European Association of Environmental and Resource Economists., Kraków, Poland.

Führ, M. and G. Röller (eds) (1991), *Participation and Litigation Rights of Environmental Associations in Europe*, Peter Lang, Frankfurt am Main.

Glasbergen, P. (ed) (1994), *Managing Environmental Disputes. Network Management as an Alternative*. Kluwer, Deventer.

Proceedings of the International Conference on Environmental Enforcement, Vol 1, Budapest, September 1992.

Jastrzębski, J. and A. Rest (1982), 'Environmental Protection in the People's Republic of Poland', *Environmental Policy and Law*, vol. 8, pp. 99-109.

Jendrośka, J (1993), 'Integrated Pollution Prevention Through Licensing Procedures in Poland', in B. Gebers and M. Robesin (eds), *Licensing Procedures for*

Industrial Plants and the Influence of EC-Directives, Peter Lang, Frankfurt am Main, pp. 99-109.

Jendrośka, J. (1991), 'Public Participation', in Z. Bochniarz and R. Bolan (eds), *Designing Institutions for Sustainable Development: A New Challenge for Poland*, Hubert H. Humphrey Institute for Public Affairs, Minneapolis, pp. 231-237.

Jendrośka, J. and J. Jerzmański (1995), 'Country Report: Poland'. In *Status of Public Participation Practices in Environmental Decision-making in Central and Eastern Europe*, Regional Environmental Center, Budapest.

Jendrośka, J. and S. King (1991), 'Environmental protection in Poland: A law in need of enforcement', *East European Business Law*, May.

Jendrośka, J. and J. Sommer (1994), 'Environmental Impact Assessment in Polish Law: The Concept, Development, and Perspectives', *Environmental Impact Assessment Review*, vol. 14 (March-May), pp. 169-194.

Jendrośka, J. and B. Wójcik (1995), 'EIA System in Poland', in *Proceedings of the International Roundtable on Practical Implementation of EIA in Central and Eastern Europe, Bratislava, Nov. 8-10, 1995.*

Levis, S.J. (1994), 'The Right to Know About Toxic Hazards and Related Trends Toward Environmental Democracy in the United States', in B. Gebers and J. Jendrośka (eds), *Environmental Control of Products and Substances*, ELNI Studies vol. 6, Peter Lang, Frankfurt am Main, pp. 143-161.

Manser, R. (1993), *Failed Transitions: The Eastern European Economy and Environment Since the Fall of Communism,* New Press, New York.

Mostert, E. (1994), *Environmental Impact Assessment in Poland. A Case Study of Expert Commissions in EIA*. Research report nr. 2, RBA Series on River Basin Administration, Delft University of Technology.

National Environmental Policy of Poland (1991), Ministry of Environmental Protection, Natural Resources and Forestry, Warsaw.

'Public Participation in Environmental Regulation in Poland: Historical Development, Existing Legislation, and Draft Proposals', report prepared by the Research Group on Environmental Law for the Environmental Law Institute, Wrocław, Poland, 1991.

Radecki, W. and J. Rotko (1991), 'Entwicklung des Natur- und Umweslschutzrechts'. In *Mittel- und Osteuropa,* Nomos, Baden-Baden, pp. 103-108.

Smith, T.T. (1992), 'Designing Enforceable Environmental Requirements — EEC', in *Proceedings of the International Conference on Environmental Enforcement*, Vol. 2., Budapest. September 1992.

Sommer, J. (1993), 'Poland', in *International Encyclopedia of Law: Environmental Law, Supplement 6.* Kluwer, Deventer.

Sommer, J. (1986), 'Regulation, Techniques and Strategies of Control and Prevention', *Environmental Policy and Law*, vol. 16, pp. 208-209.

Stec, S. (1994), 'Public Participation in CEE', in M. Nagy et al. (eds), *Manual of Public Participation*, Regional Environmental Center, Budapest.

Timberlake, L. (1981), 'Poland — The Most Polluted Country in the World?', *New Scientist,* October 22, pp. 248-250.

van Dunne, J.M. (ed.) (1993), *Environmental Contracts and Covenants: New Instruments for a Realistic Environmental Policy?* Erasmus University, Rotterdam.

van Zijst, H. (1993), 'A Change in the Culture', *The Environmental Forum*, vol. 10.

Wajda, S. and J. Sommer (1994), 'Environmental Liability in Property Transfer in Poland', in G. Goldenmann et al. (eds), *Environmental Liability and Privatization in Central and Eastern Europe*, Graham and Trotman/Martinus Nijhoff, London, pp. 179-195.

Żylicz, T. (1994), 'In Poland, It's Time for Economics', *Environmental Impact Assessment Review*, vol. 14, pp. 79-94.

6 Environmental Protection and Mass Privatization: A Management Perspective

SUSAN CUMMINGS

Introduction

Following the collapse of communism in 1989, Poland became a parliamentary democracy and initiated a process of economic restructuring designed to establish a modern market economy. Central to this process has been the large-scale conversion of enterprises and assets from state- to private-ownership through a process commonly known as 'privatization'. The process began in 1990, when the *Sejm* (Poland's Parliament) enacted the Law on Privatization of State-Owned Enterprises (Privatization Law) (1990 Dz.U. No. 51, item 298, as amended) and established a Ministry of Privatization (1990 Dz.U. No. 51, item 299) to implement the law. Since then, the number of state-owned enterprises has decreased steadily, so that by 1995 the private sector contributed 55 per cent of Poland's Gross National Product. In addition, privatization has improved profitability, investments, exports, and development strategies.

Poland has adopted a multi-track approach to privatization, providing several different methods of converting state-owned enterprises to private ownership. Over time, each privatization track have spawned variations, providing greater flexibility to meet the needs of different sectors of Polish industry and the specific needs of individual enterprises. However, the Privatization Law formally recognizes only two basic methods of privatization. First, ownership rights to a state-owned enterprise or its assets may be sold outright. Alternatively, a state-owned enterprise may be transformed into a company wholly-owned by the state treasury, with shares sold through public offerings. There are several variations on this second method of privatization, including 'sectoral privatization' of state-owned enterprises in specific industrial sectors, liquidation of the assets of smaller state-enterprises, and mass privatization through the transfer of ownership of a number of state-owned companies to National Investment Funds and the subsequent distribution of share certificates.

It was only after privatization began in Poland that participants and the Ministry became aware of the important role environmental compliance and liability issues play in the process. Potential purchasers of state-owned enterprises (especially foreign investors) were concerned with potential liability for past contamination, as well as future compliance. And environmental issues have become equally important to managers of companies and National Investment Funds since the Mass Privatization Program got off the ground in 1995.

This chapter addresses environmental issues affecting state-owned enterprises participating in Poland's Mass Privatization Program, and assesses their approach to environmental compliance and liability. In order to provide a clear context for the analysis, I begin by describing the Mass Privatization Program. Then I examine the role that environmental issues play in the program from the point of view of National Investment Fund managers.

Privatization Processes in Poland

Introduction

There are two basic methods for privatizing state-owned enterprises in Poland: liquidation and commercialization (Privatization Law, 1990 Dz. U. No. 51, item 298, art. 1). Liquidation involves the termination of the enterprise as a legal entity and the transfer of its assets to other, privately owned, legal entities. Commercialization involves the transformation of the state-owned enterprise into a state-owned company, either a joint-stock company or limited liability company, owned wholly by the State Treasury (Privatization Law, 1980 Dz.U. No. 51, item 298, arts. 3, 8 (1)). Obviously, this commercialization process does not involve the actual privatization of the company, but it is supposed to be only the first step towards the company's ultimate privatization (through mass privatization or other methods) (Privatization Law, 1980 Dz.U. No. 51, item 298, art. 2(2)).

The Inception of Mass Privatization

The idea for Poland's Mass Privatization Program (MPP) originated in 1991 (see *Rzeczpospolita*, 12 Sept. 1995). But implementation of the program was delayed, primarily for political reasons. In 1993 Poland's Parliament finally enacted a special Mass Privatization Act (1993 Dz.U. No. 44, item 202), which created National Investment Funds and procedures for mass privatization. The idea was to create a program by which a number of firms at one time could be quickly restructured and privatized. Still, it took two more years after the enactment of the 1993 Mass Privatization Act to get the program off the ground. The program only started operating in July 1995, four years after its initial inception.

Originally, more than 600 large- and medium-sized state enterprises, representing most sectors of industry, were supposed to participate in the Mass Privatization Program (MPP); companies selected for participation were listed in Council of Ministers regulations (see 1993 Dz.U. No. 78, item 368; 1993 Dz.U. No. 89, item 412; and 1994 Dz.U., No. 113, item 549). By the time the MPP was implemented in 1995, 512 former state enterprises were participating (*Rzeczpospolita*, 11 Sept. 1995). Those 512 firms were allocated among fifteen National Investment Funds (NIFs), which are joint stock companies wholly-owned by the State Treasury (1993 Dz.U. No. 44, item 202, arts. 5(1) and 3(1); 1995 Dz.U. No. 79, item 401). The express goal of the NIFs is to increase their net asset value by increasing the share values of the individual companies they hold (1993 Dz.U. No. 44, item 202, art. 4(1)).

The Main Players in Mass Privatization

The Mass Privatization Program has three main players: (1) the former state-owned enterprises privatized through MPP; (2) the NIFs; and (3) the management firms. Shares of each of the 512 MPP companies are split among four main shareholders: (1) one lead NIF gains a 33 per cent stake; (2) the 14 other NIFs share 27 per cent of the stake (1993 Dz.U. No. 44, item 202, art. 10(1)); (3) up to 15 per cent of the company's shares are distributed to the company's employees, free of charge, on the date of the transformation (1993 Dz.U. No. 44, item 202, art. 46(2)); and (4) the State Treasury retains the remaining 25 per cent of the company's shares. Following this initial distribution of company stock, the company becomes privatized as its majority owners, the NIFs (holding a 60 per cent total stake), are themselves privatized. Moreover, the NIFs can sell shares of MPP companies directly to private investors.[1] In the MPP's first year of operation, the NIFs sold shares of several MPP companies to private investors (see *Gazeta Wyborcza*, 26 July 1996). And the NIFs planned to offer shares for several companies in their portfolios on the Warsaw Stock Exchange (*Rzeczpospolita*, 18 September 1995); some MPP companies anticipated a listing on the Exchange before the end of 1996.

The second major player in the MPP, the 15 NIFs, jointly own 60 per cent of each MPP company. At the start, the NIFs were wholly-owned by the State Treasury. But as the MPP is implemented, shares in the NIFs are offered for public sale. As the NIFs are privatized, so too are the companies in their portfolios. The Mass Privatization Act provides that all 15 NIFs will be listed on the Warsaw Stock Exchange, and 85 per cent of the ownership interest in the NIFs is to be distributed to the Polish public. The State Treasury retains the remaining 15 per cent ownership interest for eventual compensation of the management firms. A General Shareholders' Meeting, Supervisory Board, and Management Board represent the interests of each NIF (1993 Dz.U. No. 44, item 202, art. 14).

Before distribution to the public, the National Investment Funds are joint stock companies, wholly-owned by the State Treasury. Each NIF initially owns a leading (33 per cent) position in 33-35 MPP companies and a minority (1.9 per cent) stake in the other 477-479 (1993 Dz.U. No. 44, item 202, art. 10(1)). As lead shareholder, the MPP plays an active management role; as a minority shareholder, it acts as a passive investor. The portfolios of the NIFs are inevitably diverse, given the large number and diversity of the 512 companies participating in mass privatization. Each NIF is responsible for designing its own strategy for managing its assets.

The third group of players involved in the MPP are the management firms, employed to manage NIFs through competitive tender offers (1993 Dz.U. No. 44, item 202, art. 21). To date, management firms have been employed to manage the assets of 14 of the 15 NIFs (*Rzeczpospolita*, 13 July 1995).[2] The management firm is a separate legal entity; its relationship to the NIF is contractual. The Mass Privatization Act anticipated that each NIF could contract with a management firm for a fixed management fee, annual performance fee, or final performance fee (1993 Dz.U. No. 44, item 202, art. 24(3)). These fees were to provide a strong incentive for management firms to maximize the long-term value of the NIF (that is, its portfolio of companies) for shareholders. The (now defunct) Ministry of Privatization expected that management firms would facilitate the restructuring, modernization, and refinancing of companies in each NIF's portfolio (see *Rzeczpospolita*, 28 September 1993). The management firm of each NIF will work with the management of those MPP companies in which the NIF is the lead shareholder to realize the NIF's investment strategy.

The relationship between a MPP company and its lead shareholding NIF is unique. the lead NIF has the right to nominate the Chairman of the company's Supervisory Board and one less than a majority of other Board members (Ministry of Privatization, 1995, p. 11). In addition, the 14 minority NIFs have the right to appoint one Board member; in practice Board members appointed by the 14 minority NIFs cooperate with the Board members appointed by the lead NIF. Meanwhile, the management board, representing the lead NIF, does not merely oversee company operations, but develops an intimate working relationship with company management.

MPP: Expectations vs. Reality

Poland's Mass Privatization Program was designed with the ultimate goal of rapidly developing MPP companies, so that they could gain access to capital markets. The players in the Program were supposed to facilitate that goal by bringing expertise and appropriate financial incentives to bear on company management. But this assumed that participating companies would begin the process in sound financial circumstances and company employees would support, implicitly or explicitly, participation in the MPP (1993 Dz.U. No. 44, item 202, art. 7(2)). Minimum qualifications for participation included: (1) $10 million in annual turnover and (2) net profits (see *Rzeczpospolita*, 18 August 1993). Each of the 512 participating companies initially met these conditions at the time they applied for inclusion in the MPP. But by the time the Program was implemented, after considerable delay, the financial circumstances of some companies had deteriorated so much that immediate assistance was required to stave off bankruptcy. The overall condition of MPP companies ranges from strong to weak, with the average company in each NIF's portfolio somewhere in the middle.

Under the circumstances, when the NIFs began operating, the first step was to conduct a sort of triage to determine which companies were in need of the quickest financial assistance. For example, NIF III successfully assisted one of its lead shareholding companies to restructure its debts, while the future of a few other of its lead shareholding companies was controlled by liquidation or bankruptcy (*Rzeczpospolita*, 29 July 1995). Similarly, NIF II, during its first year of operations, evaluated the current financial circumstances and prospects for development of each lead shareholding company in its portfolio (*Rzeczpospolita*, 30 July 1995). One of the top three goals for NIF XV was to eliminate the debts of all the lead shareholding companies in its portfolio (*Gazeta Wyborcza*, 17-18 Aug. 1996).

MPP: The Realization Phase

Lead shareholding interests in the 512 MPP companies were allocated among the 15 NIFs in two selection rounds held in July and December 1995 (see *Rzeczpospolita*, 24 July 1995 and 15 December 1995). In the first round, 413 MPP companies participated; the remaining 99 were selected in the second round. Before the drafts, the list of participating companies changed frequently; up to the last minute, companies were being added and dropped from the list, which made it difficult for the NIFs to evaluate possible selections. The NIFs drew lots to determine the order of selection in the first round. In the second round the order of selection was determined according to procedures established by regulation of the Council of Ministers (1995 Dz.U. No. 45, item 232), which operated similarly to drafts in American professional sports. Following the allocation of MPP companies, the NIFs became responsible for their day-to-day management.

The statutory goal of the NIFs is to increase the value of fund assets. There is no set method for doing this; each NIF may select its own strategy for portfolio management. However, certain approaches have become common among almost all the Funds. For example, the NIFs initially had relatively little information about the MPP companies allocated to them. Before the allocations, representatives of each NIF visited and/or researched all the MPP companies, but the large number of participating companies made it impossible for any NIF to evaluate all of them in detail (see *Gazeta Banco*, 27 August 1995 and *Now Europa*, 23 August 1995). Consequently, the first step almost every NIF took following the allocation was to conduct a detailed assessment of their lead shareholding companies. For most Funds, these assessments took up the entire first year of MPP's realization phase.

There is general agreement that the Funds should work to consolidate minority shareholding interests. In practice, the structure of minority shareholding under the MPP has become cumbersome. So the NIFs are working together to reallocate shares so that each Fund would hold more shares in fewer companies; they would no longer own some part of every single MPP company. This re-allocation was to occur in at the beginning of October 1996 (*Rzeczpospolita*, 21 August 1996). Some Funds have even expressed interest in becoming the lead shareholders of companies in which they currently hold minority stakes (see *Rzeczpospolita*, 30 August 1995; *Gazeta Wyborcza*, 26 July 1995).

MPP: Environmental Management

Each NIF's approach to environmental protection falls within its discretionary management strategy. Neither the Mass Privatization Act nor any other statute imposes special environmental protection requirements on Fund managers. However, the MPP companies remain subject to environmental regulation, just like any other entities doing business in Poland. Further, even though the State Treasury retains an ownership interest in MPP companies, the government does not assume their environmental liabilities (Ministry of Privatization, November 1995, p. 16). Neither are the NIFs liable for the environmental problems of companies in their portfolios — only the companies themselves are responsible for the costs of environmental compliance and liabilities — although company liabilities may significantly affect the value of NIF assets.

This provides the NIFs with some incentive to address environmental issues affecting their lead shareholding companies. Environmental risks become, in effect, part of the overall financial risk the NIFs must manage. But, as with the rest of its management strategy, each NIF is free to develop its own approach to environmental protection. It is to be expected that the form of environmental management will vary from one Fund to the next. Ultimately, the environmental performance of each MPP company will affect its ability to obtain loans and financing, attract strategic investors and partners, and remain competitive in increasingly eco-conscious markets.

Polish Environmental Regulation: The Company's Perspective

Environmental Regulation of MPP Companies

Polish environmental legislation controls and regulates the use of the environment by all economic actors, including MPP companies. Environmental costs can be classified into three groups: (1) compliance costs; (2) operating costs, including fees and fines; and (3) liabilities. These are among the costs of doing business in Poland. As such, NIFs and MPP companies must develop cost-effective strategies for dealing with them.

For industrial companies, the use of environmental resources in the first place requires permission from the competent body of state administration. In general, permissions, usually in the form of a permit or administrative decision, specify the kind, extent, and duration of activities for which they are granted. Whether permitted or not, the MPP company that uses environmental resources is subject to use fees, according to a fee schedule set by law. Companies that use the environment without a permit or fail to comply with permit conditions are in violation of the law, and may be fined or shut down. Thus, an MPP company's first environmental obligation is to gain any necessary administrative permissions.

Current Polish environmental legislation focuses on air pollution, water pollution and waste (see 1980 Dz.U. No. 3, item 6, as amended; and 1974 Dz.U. No. 38, item 240, as amended). Starting with air pollution, any plant that emits pollutants into the atmosphere must obtain a permit specifying how much pollution it can emit (1980 Dz.U. No. 3, item 6, art. 30). The permit is granted for a fixed period, and specifies the types and amounts of pollutants that may legally be emitted, both from the plant as a whole and from each discrete source within the plant. In addition, permits specify various obligations relating to monitoring and reporting of emissions. Unlike in the United States, legally permitted pollution emissions are not free of charge in Poland. Emissions within permitted levels are subject to per unit fees collected by the local governmental authorities (See 1993 Dz.U. No. 133, item 638, as amended, art. 2). And fines are imposed for all emissions that exceed permitted levels (1995 Dz.U. No. 79, item 399, art. 1). Typically, these fines are double the normal fees. Ironically, despite the legal obligation to obtain a pollution permit, companies that pollute without a permit are not deemed to be in violation of the law, and therefore are not subject to fines.

Water pollution permits are required for activities involving water consumption, waste water discharges, and the operation of water facilities (1974 Dz.U. No. 38, item 240, art. 20). In addition, water users must pay fees for permitted (that is, legally authorized) uses (1993 Dz. U. No. 133, item 637) and for unauthorized uses (1974 Dz.U. No. 38, item 240, art. 26; 1995 Dz.U. No. 79, item 400). Water use permits, which may also contain conditions pertaining to the types and amounts of permitted waste, are valid for a fixed period, set on a case-by-case basis. The local authorities collect assessed fees for water use and impose fines for permit violations.

With respect to industrial solid waste disposal, Poland has not yet implemented a recently enacted comprehensive regulatory regime, although non-recyclable solid wastes must be disposed of at sites designated by the competent authorities (according to specified procedures). Waste disposal is subject to a per unit charge (1980 Dz.U. No. 3, item 6, art. 54; 1993 Dz.U. No. 133, item 638, art. 2; 1980 Dz.U. No. 24, item 91), and fines are imposed for illegal waste storage and for failure to fulfil the conditions of a decision on the siting of a waste storage facility (1993 Dz.U. No. 133, item 638, art. 7).

In addition to air use and pollution, water use and pollution, and waste disposal, Polish law also regulates the generation of noise. Although noise emissions do not require an administrative permit, and no fees are charged for noises, local administrators may set a noise limit for a facility if it emits noise in excess of generally applicable regulations (1980 Dz.U. No. 3, item 6, arts. 49-52). Once an individual noise limit is set, a fine may be imposed for exceeding that limit (1993 Dz.U. No. 133, item 638, art. 2).

The NIFs' Approach to Environmental Compliance

The NIFs have designed a three-step program for the general management of their respective portfolios: (1) assessment; (2) creation of a strategy; and (3) implementation (Seminar on 'Polish Mass Privatization National Investment Fund Program', 20 June 1996). Environmental issues will be treated within this structure. In other respects, each individual Fund defines its own approach to the specific environmental issues of its lead shareholding MPP companies. Among 500 MPP companies there is sure to be great variation in the type and scope of environmental issues. So it is to be expected that the NIFs will take various approaches to dealing with different companies and different environmental issues. Two extreme approaches to environmental protection may help to define the range within which each Fund's actual approach(es) will fall. One extreme, similar to the way state enterprises in People's Poland approached environmental protection, is a passive policy. A Fund taking this approach might leave questions of environmental policy to the discretion of the MPP company's management. The management would decide how much (if at all) to comply with environmental protection requirements, including permitting rules. Alternatively, a Fund might take a pro-active approach to environmental protection, imposing specific obligations and targets on all lead shareholding companies in its portfolio.

During the initial assessment phase of NIF management, the Funds generally have taken an approach closer to the extreme passive role, though this may change as the MPP progresses. Once a Fund has defined an overall approach to environmental protection issues, it will, within that framework, develop strategies tailored to the specific needs of each of its lead shareholding companies. For example, a Fund preparing an MPP company for sale to a strategic investor will need to conduct a precise audit of environmental compliance and potential liabilities, but will not be as concerned with issues relating to future compliance; those will become the responsibility of the investor following the sale. For another company, which the Fund expects to retain, long-term compliance issues are likely to be the focus.

A Case Study in NIF Environmental Management

The approach of one NIF, Hetman National Investment Fund (Hetman NIF), exemplifies how NIFs can approach the environmental management of portfolio companies. Hetman NIF began by conducting environmental audits of each lead shareholding company in its portfolio, as part of its overall company assessment. The Fund conducted this audit internally, to avoid the costs of hiring outside consultants. First, it prepared a questionnaire on environmental issues, which it sent to each MPP company. The questionnaire addressed issues ranging from production technology, regulatory compliance, environmental management, potential risks, and investment requirements. These questions were media-specific to assess actual and potential environmental liabilities for air pollution, water pollution, waste disposal, and soil and groundwater contamination. The questionnaire also addressed basic land-use planning and facility siting issues. And it requested that each company provide the Fund with relevant permits, licenses, and certificates, as well as the amount spent on environmental fees and fines. The completed questionnaires became the basis for Hetman NIF's assessment of environmental compliance and liability issues for each of its lead shareholding companies. If further information is needed to complete the initial environmental assessment, the management firm may conduct site inspections or require due diligence investigations.

The Fund, having prepared an initial environmental assessment of each lead shareholding company, may use this information in a couple of ways. First, if the NIF anticipates the sale of shares in a given MPP company, the environmental assessment may provide useful information for negotiations. Second, for companies in which the Fund intends to maintain a long-term ownership interest, the environmental information may provide the basis for an active environmental management program.

If the Fund decides to actively manage its lead shareholding companies' environmental compliance, it may prepare a strategy for each individual company, focusing on their specific environmental issues, as revealed by the completed audit. The overall management goal would be to: (1) promote compliance with existing environmental regulations; (2) ensure the company's ability to comply with stricter environmental regulations (which are anticipated as Poland endeavors to enter the European Union); (3) where cost-effective, further reduce the environmental impacts of production. This strategy might involve the preparation of Environmental Guidelines for each lead shareholding company. The Guidelines would address basic environmental issues, such as operational compliance, risk management, and liabilities, and performance beyond mere compliance. The Fund may also encourage the company management to develop an Environmental Action Plan to address its operational needs.

The Creation of Environmental Protection Strategies

Given the weak financial condition of many companies participating in the MPP, any environmental protection efforts must be cost-effective. Inevitably, Funds and MPP companies will apply a cost-benefit analysis to determine the feasibility of environmental investments and other activities to reduce pollution emissions and avert liability. Indeed, economic performance concerns may decide environmental performance issues. For example, companies can be expected to work toward compliance with current legal requirements if the costs of pollution reduction are exceeded by the expected reduction in environmental fees and fines. In this way, environmental issues will represent one element of the company's overall strategy for improving its financial condition. But this may actually bode well for environmental protection at MPP companies.

The environmental problems of many MPP companies are largely symptoms of inefficient production. By reducing those inefficiencies, the companies would automatically improve their environmental and economic conditions at the same time. In other words, the overall strategy to improve the financial condition of MPP companies may bring substantial environmental benefits even in the absence of a specific environmental protection strategy. It is important to note, however, that the MPP does not provide financing for environmental or other investments. General economic or environmental investments must be self-financed, either with the company's own resources or through resources provided by the capital market. A NIF may actively encourage compliance and work to stimulate increased environmental awareness at its lead shareholding companies, but the company management is ultimately responsible for realizing environmental and other investments. The Funds may, of course, assist their portfolio companies in identifying funding sources and options, including for environmental investments.

After the initial environmental audit and the development of a strategy for improving compliance by lead shareholding companies, an NIF may institute a monitoring and reporting regime. This would not only provide the Fund with regular information on the environmental performance of companies in its portfolio, but would keep company management attention on environmental issues. A company that is attending to its environmental performance is more likely to exploit cost-effective gains from additional environmental investments. In addition, both the Fund and the company would be better positioned to deal cooperatively with the environmental authorities, possibly to the benefit of the company and Fund. They would also be better position to respond to future changes in environmental regulations.

Conclusion

The main goals of Poland's Mass Privatization Program are to restructure and ultimately transfer ownership of the 512 MPP companies into private hands. After allocation to the National Investment Funds, the MPP companies are managed according to market criteria. The main goal of the NIFs, at this stage, is to improve their economic condition and increase their asset value. Environmental protection is an inseparable element of this effort. The NIFs certainly have the management authority, and should have the incentive, to actively manage their lead shareholding companies to improve environmental performance in cost-effective ways. Whether or not an individual Fund will stress pro-active environmental management will depend, of course, on its calculations of the costs and benefits of specific environmental investments. Given the unique circumstances of the 512 Polish companies participating in

MPP, especially their potential for improved productive efficiency, the program provides a useful mechanism for fostering environmental protection while realizing the program's ultimate economic goals.

Notes

1. Private investors who want to acquire a majority holding will have to deal with several parties because of the initial distribution of shares in the MPP. They may purchase up to 33 per cent from the lead NIF, a total of 27 per cent from the other 14 NIFs, and up to 15 per cent from the company's employees, who received their shares free of charge.
2. The 14 current management firms are:

I	NIF	BRE/IB Austria Management Sp. z o.o.
II	NIF	Hevelius Management Sp. z o.o.
III	NIF	Trinity Management Sp. z o.o.
IV	NIF	Konsorcjum Faiffeisen Atkins
V	NIF	PTP-Kleinwort Benson
VI	NIF	Chase Gemina Polska Sp. z o.o.
VII	NIF	Kazimierz Wielki Fund Management Polska Sp. z o.o.
VIII	NIF	KP Konsorcjum Sp. z o.o.
IX	NIF	Fidea Management Sp. z o.o.
X	NIF	KN Wasserstein Sp. z o.o.
XI	NIF	Eurofund Management
XII	NIF	Yamaichi Regent ABC Polska Sp. z o.o.
XIII	NIF	Westfund Polska Sp. z o.o.
XIV	NIF	Creditanstalt SCG Fund Management

References

1974 Dz.U. No. 38, item 240. *Ustawa z dnia 24 października 1974 r. — Prawo wodne.*

1980 Dz.U. No. 3, item 6. *Ustawa z dnia 31 stycznia 1980 r. o ochronie I kształowaniu środowiska.*

1980 Dz.U. No. 24, item 91. *Rozporządzenie Rady Ministrów z dnia 30 września 1980 r. w sprawie ochrony środowiska przed odpadami I innymi zanieczyszdzeniami oraz utrzymania czystości I porządku w miastach I wsiach.*

1990 Dz.U. No. 51, item 298. *Ustawa z dnia 13 lipca 1990 r. o prywatyzacji przedsiębiorstw Państwowych.*

1990 Dz.U. No. 51, item 299. *Ustawa z dnia 13 lipca 1990 r. o utworzeniu urzędu Ministra Przekształceń Własnościowych.*

1992 Dz.U. No. 133, item 638. *Rozporządzenie Rady Ministrów z dnia 23 grudnia 1992 r. w sprawie opłat za gospodarcze korzystanie ze środowiska I wprowadzanie w nim zmian.*

1993 Dz.U. No. 44, item 202. *Ustawa z dnia 30 kwietnia 1993 r. o narodowych funduszach inwestycyjnych I ich prywatyzacji.*

1993 Dz.U. No. 78, item 368. *Rozporządzenie Rady Ministrów z dnia 18 sierpnia 1993 w sprawie określenia jednoosobowych spółek Skarbu Państwa w celu wniesienia ich akcji lub udziałów do narodowych funduszy inwestycyjnych.*

1993 Dz.U. No. 89, item 412. *Rozporządzenie Rady Ministrów z dnia 14 września 1993 w sprawie określenia jednoosobowych spółek Skarbu Państwa oraz przedsiębiorstw państwowych, które zostaną przekształcone w spółki w celu wniesienia ich akcji lub udziałów do narodowych funduszy inwestycyjnych.*

1993 Dz.U. No. 133, item 637. *Rozporządzenie Rady Ministrów z dnia 27 grudnia 1993 r. w sprawie opłat za szczególne korzystanie z wód I urządzeń wodnych.*

1994 Dz.U. No. 113, item 549. *Rozporządzenie Rady Ministrów z dnia 18 października 1994 zmieniające rozporządzenie w sprawie określenia jednoosobowych spółek Skarbu Państwa w celu wniesienia ich akcji lub udziałów do narodowych funduszy inwestycyjnych oraz rozporządzenie w sprawie określenia jednoosobowych spółek Skarbu Państwa oraz przedsiębiorstw państwowych, które zostaną przekształcone w spółki w celu wniesienia ich akcji lub udziałów do narodowych funduszy inwestcyjnych.*

1995 Dz.U. No. 45, item 232. *Rozporządzenie Rady Ministrów z dnia 11 kwietnia 1995 w sprawie określenia trybu wnoszenia akcji jednoosobowych spółek Skarbu Państwa do narodowych funduszy inwestycyjnych.*

1995 Dz.U. No. 79, item 399. *Rozporządzenie Rady Ministrów z dnia 20 czerwca 1995 w sprawie wysokości, zasad I trybu nakładania kar pieniężnych za nieprzestrzeganie wymagań ochrony środowiska oraz współczynników różnicujących wysokość kar pieniężnych.*

1995 Dz.U. No. 79, item 400. *Rozporządzenie Rady Ministrów z dnia 20 września 1995 r. w sprawie kar pieniężnych za naruszanie warunków, jakim powinny odpowiadać ścieki wprowadzane do wód lub do ziemi oraz współczynników różnicujących wysokość kar pieniężnych.*

1995 Dz.U. No. 79, item 401. *Rozporządzenie Rady Ministrów z dnia 20 czerwca 1995 w sprawie określenia narodowych funduszy inwestycyjnych, do których zostaną wniesione akcje jednoosobowych spółek Skarbu Państwa.*

Ministry of Privatization (1995), *Information relating to the Universal Share Certificate.* Privatization Ministry, Warsaw.

Seminar on 'Polish Mass Privatization National Investment Fund Program', 20 June 1996, London.

7 Polish Greens and Politics: A Social Movement in a Time of Transformation

PIOTR GLIŃSKI

The Ecological Movement in the 1980s

During the communist era (at least until 1980), there were no truly independent environmental organizations in Poland. A few 'official' conservation associations were permitted within the system and played by its rules. They were fully controlled politically and their role in the system was defined by the authorities.

It was only in the 1980s that independent, spontaneous, and collective pro-ecological groups developed on a large scale in Poland. Born as a typical grassroots protest movement, the ecological movement sought specific social changes. A particularly important factor sparking and shaping the development of the ecological movement was the creation, growth, and dramatic crack-down on official activity of the independent trade union, Solidarity. The explosion of genuinely public activity during Solidarity's 16 months of legal operation bore fruit in the form of various social initiatives, including ecological initiatives.

The short history of the Polish ecological movement divides into two developmental phases which reflect the dominance of specific organizational forms as well as different social and political circumstances. The first phase coincided with the creation of Solidarity in 1980, while the political changes of 1989 launched the second. The political conditions of social resistance to the Communist authorities were imprinted on the character of the first phase. For the environmental movement, it was a time of demonstrations and protest. Greens more than any other group engaged in symbolic acts intended to appeal to emotions more than to rational choices and strategies. Typical of this period were spontaneous informal organizations and groups, often arising in response to local conflicts over environmental dangers. These were the first independent citizens' initiatives in Poland that sought to pressure the authorities from below concerning ecological problems. There was a very low level of formal organization among these groups, which possessed virtually no organizational or negotiating skills. Because there was little coordination among groups, acquiring and exchanging information was very difficult.

The attempted liberalizations of the Communist system in Poland in 1987 and 1988 allowed the first authentic nongovernmental ecological organizations to emerge. Because of the character of the totalitarian system, this independent civil activity had the nature of resistance movements and represented enclaves of oppositional civil society.

By the beginning of the transformation period in 1989, Poland boasted a relatively broad and diverse ecological movement that encompassed more than one hundred organizations and groups, and more than a dozen ecological foundations (altogether 135 independent, formal and informal, environmental organizations) (Gliński, 1996, pp. 155-6). The movement's fundamental trait was and still is a pluralism bringing together various segments of society: young people, religious groups, technical specialists, professionals, local associations, etc. Individual groups differ in their degree of institutionalization, demands, legal status, involvement in societal conflicts, the geographic and social range of their activity, involvement in politics, and their search for alternative cultural models. All, however, share a rejection of all forms of violence. At least until mid-1989, many ecological groups and organizations, and in particular those organized among young people, experienced political harassment. We should note that the ecological movement primarily attracts members of the intelligentsia and young people. Farmers are less in evidence, while workers are conspicuously under-represented.

Two general orientations were and still are present in the movement as a whole. One continues the traditional 'preservationist' approach to ecological problems, striving to protect nature from the environmental degradation caused by human beings and contemporary civilization. This tendency is frequently characterized by its critics as the 'ecology of fear'. An alternative orientation characterizes groups searching for new cultural values and striving to change the fundamental 'civilizational paradigm'. These groups draw on counterculture traditions that challenge consumerism and economist. They emphasize humanity's unity with nature, which entails the need for people to renounce their domination over the environment. Such an 'introspective' approach to ecology is congruent with the principles of 'deep ecology' (Devall and Sessions, 1985).

The Development of the Green Movement During the Transition Period

The second phase of the environmental movement, which began in 1989, witnessed both quantitative and qualitative changes in the movement. The political changes that took place in Poland during the summer of 1989 caused a dynamic quantitative growth in the number of environmental organizations. The period of the most dramatic growth was in 1991 and 1992. By 1995 there were around 700 environmental organizations in Poland (520 associations and informal groups, and 180 foundations).[1]

The internal structure of the movement became more diverse after 1989. Because of rather vague group identities, it is more difficult now than it was a few years ago to build a typology of participants or organizations in the movement. Generally speaking, at least fourteen groups of participants can be distinguished within the Polish environmental movement in the 1990s. They include:

- The Polish Ecological Club
- The League for Nature Protection and other post-systemic organizations
- The Green Federation
- Animal rights activists and the vegetarian movement
- Other youth organizations (students and pupils, scouts, the remnants of the 'I Prefer to Be' Movement and the 'Freedom and Peace' movement, etc.
- 'Deep ecology' circles
- Conservationists
- Local organizations
- Educational organizations
- Specialists and experts (economists, lawyers, organizations of ecological farmers, alternative energy and technology groups, etc.)
- Religious and quasi-religious groups (Catholic, Buddhist, Hindus, etc.)
- Artistic and para-artistic milieus
- Service and support institutions
- Plus various other quasi-ecological countercultural groups, quasi-nongovernmental organizations, certain foundations, noted individuals, etc.

The main fields of the movement's activity in this period covers some two dozen problems and campaigns, including:

- Anti-nuclear actions (mainly in 1989, opposing the construction of the Żarnowiec nuclear power station)
- Opposition to hydro-investments (such as Czorsztyn dam)
- Recycling and waste management campaigns
- The 'Wild Life' campaign

- Traditional forms of conservation and nature protection
- Defense of animal rights
- The 'Green Lungs of Poland' project
- Opposition to transport pollution and the promotion of bicycles
- The 'Vistula Now!' campaign
- The Carpathian Euro-region campaign
- Local defensive actions
- Ecological education
- Strategic cooperation with local self-governments
- Development of ecologically friendly agriculture and food production
- Combating environment and health problems
- Addressing environmental policy and environmental action plan issues
- Protection of the Baltic Sea
- Fostering the growth of environmental NGOs and banks
- Involvement with Greenpeace International
- Trying to bring about a general change in the attitude of people towards nature (for instance, by promoting the perspective of deep ecology)
- Promoting the institutional development of the movement
- Actions at the border of the movement such as 'Cleaning the World' action and the 'Animals' TV broadcast.

Although members of the environmental movement do not all follow the same ideology, the processes of 'maturation' and 'self-education' are leading to the convergence of collective beliefs and common thinking.

The environmental movement in Poland is a community bound by only the most general goals. That specific groups often have very different aims is a natural result of specialization and the free choice groups have in determining their objectives. However, a social movement must be able to recognize a common purpose in order to function. Despite the ideologically radical programs of many pro-environmental groups, it can be said that the idea of environmentally appropriate development (often called 'sustainable development' or 'eco-development') serves as a common goal of the movement. They also share a long list of secondary goals of a more symbolic nature, such as: opposition to nuclear energy, opposition to large dam construction, the preservation of ecologically significant habitats and unique ecosystems, energy conservation, the introduction of waste management systems, and so on. The problem for the movement is determining priorities among these goals.

The social composition of the membership in the Polish environmental movement only partially reflects the societal structure of most Western new social movements. The basic difference is the absence of a large middle class in Poland, especially the lack of a 'new' middle class. As a result, we see the development of a 'new social movement' without a middle class, which in the Polish context automatically limits its scale to a 'non-mass' movement. The role of the middle class is, in part, assumed by the intelligentsia[2] in the Polish environmental movement. The involvement of the intellegentsia entails a more rational, pragmatic and self-limiting character of the movement. While the ethos of the future Polish middle class will in large part shape the future of the movement, the former middle class plays no role in the movement. As in the West, youth play a fundamental role in the Polish Greens, especially high school and university students. The movement's counterculture background is important, as in the West, for it is from the counterculture that many of its participants emerged and its networks were established. The list of other traits in common with the West includes the urban character of the movement, the predominance of highly educated participants, and the conspicuous absence of the working class, *nouveau riche*, or political elites.

Besides the quantitative changes in the composition of the movement, the Polish nongovernmental organization (NGO) sector is also characterized by qualitative changes in the form and style of organizations' activities. In the 1980s and in the initial stages of transformation, nongovernmental organizations primarily engaged in resistance activities directed against the state administration. They tended to employ the simplest forms of social group action, staging protests and interventions, and articulating the interests of particular communities or groups. Dominant forms of activity today are more typically civil: monitoring the behavior of those in power, participating in the processes of public policy preparation and implementation by cooperating with local administrators, pointing out problems to be solved, suggesting legal solutions, and lobbying; and taking the initiative in non-state social spheres, sometimes through direct action, sometimes by preparing alternative solutions of social problems.

Participants in the Polish environmental movement engage in 'new' types of activities and methods of operating, typical of new social movements. In the Polish context, these 'new' elements include using humor of the absurd, distancing oneself from reality, and rejecting all forms of violence. The movement is characterized typically by 'new' organizational forms. Lateral and non-hierarchic structures, loose networks, and a low level of formality dominate. Fundamental features include direct democratic governance, consensus, and the close interaction of members.

At the same time, environmental organizations are developing more professional structures, similar to the structures found among new social movement groups in the West. Changes in the movement's structure and methods of activity reflect a gradual maturation and professionalization. Rational strategies increasingly guided its activities. More and more, protest actions were supplemented or even replaced by constructive and positive action. Informational networks were set up and nongovernmental experts engaged. Movement activists trained themselves to conduct environmental campaigns and programs, to administer nongovernmental organizations, and to understand legal codes. They learned how to work with state agencies and the mass media and to raise money for ecological activity.

The 'maturation' of the Polish environmental movement, characteristic of the entire Polish nongovernmental sector, depends primarily on the following factors:

- changes in the forms of organizations' activity that were discussed above;
- growth of an information and communication network both within and outside the sector, including the creation of regional and countrywide cooperative institutions such as the Polish Foundations Forum and the Regional Nongovernmental Initiative, and the growth of the third sector press;
- development of contacts within the local milieu, especially with local government, regional groups, and academic groups, plus the beginnings of cooperation with the mass media (Wygnański, 1995a) and the parliament;
- development of the third sector services institutions such as the Information Bank of Nongovernmental Organizations KLON/JAWOR, the Service Bureau of Self-Help Initiatives, and the Service Office for Environmental Movement; and
- professionalization of nongovernmental organizations' activities — as in Western new social movements, this has been characterized by the employment of professional organization leaders, the accumulation of resources from outside the sector, a relatively inert member body, aspirations to represent a broader public group interest, and attempts at public lobbying (Zald and McCarthy, 1990).

Although the maturation process of the third sector in Poland is still in its initial phase — for example, NGOs show relatively little interest in the legislative process — the trend of the changes is clear.

The immediate reasons and causal mechanisms of these changes are processes of self-education (Eder, 1993) of the participants of the third sector, and the influence and help of the western partners.

Self-education can be illustrated by the history of Czorsztyn dam protest. From 1988 to 1993, the ecological movement was involved primarily in two large protest campaigns, one against the construction of a nuclear power plant in Żarnowiec and the other against the construction of a dam at Czorsztyn. The movement employed tactics such as mass media campaigns, road blockades, demonstrations, and even hunger strikes to achieve success in the first case. Work on the power plant was halted. The main reason for the Polish government's decision, however, appears to have been economic: it lacked the money to pay for continued construction. The protests against the Czorsztyn dam took a similar form. In this case, however, annual summer blockades of the approach roads to the site and propaganda campaigns in the mass media did not produce the same practical results. The protests did not prevent construction of the dam, though they did force the authorities to invest in certain ecological installations (mainly sewage treatment plants) and lent notoriety to the ecological problems caused by dams.

In 1991 some environmental groups added other types of activity to their repertoire: direct discussion and contact with the Ministry of Environmental Protection, lobbying members of parliament, mobilizing and activating a broad coalition of nongovernmental environmental organizations (including experts) against the dam, and, finally, generating a report by independent experts of the circumstances and effects of constructing the dam. These tactics were opposed by some radical youth groups (in particular, Freedom and Peace). Despite sharp conflicts within the movement over the new activities, they marked a new level of cooperation with the government and parliament, and increased intra-movement coordination.

As a result of steady pressure exerted by ecology groups combined with favorable personnel changes in the Ministry of Environmental Protection, contacts between the movement and the Ministry intensified in 1992. The Ministry established an office that sponsored monthly meetings between representatives of the Ministry and participants in the movement. The goal of these meetings was to make it easier for ecologists to bring grievances to the attention of the minister on a regular basis, and to exchange information regarding concrete ecological problems and the state's environmental policies. On the other hand, the government meant the meetings to mobilize the environmental movement in support of the Ministry's policies. Given the proper organizational preparation, working together in this way was intended to lead to permanent forms of participation by nongovernmental organizations in the ecological decision-making process.

This process of 'politicization' brought with it efforts to change the ecological movement's internal structure. The Polish ecological movement moved gradually into the second phase of development, the phase of constructing effective, professional nongovernmental organizations. Many ecological foundations and groups decided to employ a professional staff in their search for financial support for their activities (often from foreign sources). Ecological organizations that were no more than Communist 'fronts' before 1989 began to function independently. The movement also started to construct an internal information network. Efforts were undertaken to coordinate the activities of different groups and organizations. Most ecological groups participated actively in all of these processes. Attempts at coordination can be traced back to 1988, but it was not until September 1991 that a Service Office for the Ecological Movement was set up, thanks to aid provided by Western sponsors.

Foreign assistance played a significant role in developing the environmental movement in Poland. From 1990 to 1994 foreign foundations and ecological funds allocated at least 1.6 million dollars to Polish pro-ecological organizations. (Gliński, 1996). The extent of foreign aid to the third sector in Poland is in reality much greater because a part of Polish government grants to environmental NGOs comes from foreign aid, principally from the EU's PHARE program. Many voluntary organizations, especially Polish nongovernmental institutions supporting the growth of the environmental third sector, are directly financed by PHARE and USAID. Foreign aid was especially important to NGOs in the first phases of organizational growth when there were practically no other sources of financing. The private sector was very weak and the people who were creating the third sector in a post-Communist Poland lacked any management skills or financial know-how.

A very large, and often unappreciated, role in the growth of the nongovernmental sector in Poland is also played by non-financial forms of foreign aid, indirectly tied to the activities of western foundations and aid organizations such as the American Peace Corps. This role includes:

- providing professional organizational skills;
- inspiring a move toward a more professional attitude and development of a commitment to volunteer work by building a positive image of a volunteer; and
- forcing the Polish governmental organizations to interest themselves in the local third sector problems in order to promote environmental protection and the growth of civil society in Poland.

These changes did not come about without resistance, principally of two kinds. In the first place, it is very difficult to overcome ingrained habits of falling back on easy ways of doing things, for instance conducting protest type actions which do not require special skills. This happens especially in conditions of new political freedom, when 'courage gets cheaper'. It also happens when governing political elites show a largely negative attitude toward the third sector while the social prestige of the state and political classes (as is shown by public opinion polls) is still low. This explains why the positive changes of environmental organizations' activity which was discussed above came to pass as if outside the sphere of immediate contact with the world of politics. Studies of the third sector in Poland reveal that the predominant type of relation between NGOs and the state administration is simply a lack of contact; between NGOs and political parties and politicians, it is a lack of contact coupled with a complete unwillingness even to initiate contacts (Wygnański, 1995b).

Second, during a time of systemic transformations characterized by weak democratic institutions, subtler civil forms of activity based on nongovernmental organizations' commitment and competence sometimes tend to weaken their political position and to erode their power. Primitive, demagogic protest actions are often more effective in furthering the interests of some groups. Polish environmental organizations thus often have to face the classic dilemma of force vs. competence (Wygnański, 1995a).

Despite the movement's functioning information network and the coordination and professionalization of some groups, the environmental movement in Poland was still beset by numerous organizational shortcomings that limited its effectiveness. These included: an inability to devise action strategies for ecological campaigns and above all for campaigns to mobilize public support and recruit new members; the preponderance of defensive and protest-oriented activity; shortages of funds, professional personnel and expert advisors (in the field of environmental law and economics); and a lack of basic negotiating skills.

The resources available to the ecological movement in Poland have undergone remarkable changes during the period of transformation. At the beginning of the process, the movement lacked almost all the resources typical of social movements: it had no funds or fund-raising abilities, organizational structures, organizational and managerial skills and knowledge, membership, etc. Most of these resources have become accessible for the movement during its professionalization and maturation, except for skilled leaders and membership. There is still a shortage of recognized leaders of high authority who can elevate the movement's level of coordination and effectiveness. Membership is still limited in number as a result of the general passivity of society and inadequate efforts by activists to increase the base of their organizations, which in turn is a result of insufficient organizational skill and a particular reluctance to engage in mass mobilization or recruitment. This attitude is a reaction to the communist era, when any public mobilization campaign was perceived as artificial and hypocritical.

The Refusal of Political Participation

Although 1989 offered environmental activists opportunities for participation in new political structures and institutions, not many members of the ecological movement involved themselves directly in these changes. Some members of 'Freedom and Peace' became parliamentarians and state officials, and some of these continued their ecological activity, creating the 'Ecological Faction' in the Democratic Union, the strongest new political party.[3] Their initiatives had little impact, above all because of the lack of interest on the part of the Democratic Union's political leadership.

Generally speaking, the ecological movement has for the most part rejected participation in the new, official political structures. Some important groups such as the Green Federation attempted political action as early as 1988, but quickly abandoned it as contrary to their fundamental values. The Federation's organizational rule of thumb is a minimum of bureaucracy and formalized structure. It also emphasizes advocacy over governance. Member groups delegate representatives to the Federation's coordinating body, which confines itself to two tasks: exchanging information and coordinating activities above the local level (while according full respect to the principle of local autonomy). Given these views of its adherents, the Green Federation's stress on ecology over politics clearly differentiates it from the Polish Green Party.

While a majority of ecological groups adopted a neutral position towards the changes taking place in Poland in 1989, other behaved as if no change of power had taken place whatsoever. For example, 'Freedom and Peace' and anarchist groups organized a demonstration against Tadeusz Mazowiecki's government a mere two weeks after he formed his cabinet.

What motivated this rejection of normal political structures? Why did Polish Greens hold back from active and direct participation in the process of political transformation, despite the collapse of the repressive system and new possibilities for free activity in public life? The most general reasons for this have to do with material conditions and certain traditional values. This complex of causes not only impedes development of political activity among the movement's participants, but, more importantly, weakens the ecological organization's overall impact, its effectiveness, size, and ability to mobilize social support.

The need to earn money in new, difficult economic circumstances has forced many, including those inclined to take leadership roles in social and political movements, to divert increasing time and energy away from activist careers. The threat of unemployment, deterioration of social services received from the state and from enterprises, and the need to compete on the job market have led many prominent participants in the ecological movement to cease or curtail activities.

The second fundamental barrier to political activity among ecological activists is a cultural heritage of real socialism. Years of dramatic sociopolitical change have left Polish society without a cohesive, stable system of values and social norms, and a practical culture for developing independent social movements has yet to coalesce. In the past, social movements and nongovernmental organizations have emerged only in societal enclaves. As exceptional phenomena they lacked mass appeal and mass support, particularly in the context of broader, society-wide transformation.

Reluctance to engage in civic, social, and political activity is embedded in popular, one might even say folkloric, behaviors. Concern over status often prevents compromise in societal conflicts, organizational efficiency is not a high priority, and traditions of 'oppositionism' foster an aversion to formalized activity. Some of these 'inherited' cultural models are a legacy of real socialism. Passivity and 'learned helplessness' (the refusal to take initiative and bear individual responsibility for actions, the expectation that the totalitarian state will act as provider) were basic components of the real-socialist world-view. Social and political engagement has also been diminished by the effects of rapid transformation on individuals' capacities for understanding the social world and interpersonal relations. The past and present continually are being redefined, while prospects for the future are unclear. Because value systems have been shifting, along with individual, group and social identities, there are no effective rules of behavior adequate for the present situation (Kolarska-Bobińska, 1990).

Lack of participation of environmental groups in the process of political transformation can also be explained by characteristics of the Polish environmental movement directly related to these cultural conditions. Above all, people in the ecological movement (especially young participants) are alienated from political institutions. They regard politics as an unethical and deceitful activity, one in which the satisfaction of individual ambitions is more important than general societal interests. This attitude was formed under real socialism and was automatically carried forward into the period of systemic change.

The fact that ecological groups organized themselves around a community of values rather than interests has also been a factor in their rejection of politics. In the eyes of movement activists, commitment to counterculture values contradicted the very idea of a political career that requires forming coalitions and accepting compromises. A further reason for the rejection of the world of politics lay in the dominant forms of activity characteristic of ecological groups. During the period of Communist rule, they became accustomed to protest and to destructive activity, which they considered especially attractive forms of mobilization and resistance. The appeal of these activities increased greatly after 1989, when the cost of civil courage depreciated and 'direct action' no longer provoked serious repression. But real political activity, the erection of new structures and political institutions, demanded a reorientation in mentality toward constructive and creative action. Polish environmental activists were not prepared for such a profound shift in attitude. A peculiar 'veterans' syndrome' surfaced: more than making activists incapable of change, old habits made them reject the new circumstances altogether. They pulled back from the external institutional world at the cost of ignoring new possibilities. As far as ecological questions were concerned, activists continued to be interested mainly in what they could rebel against, that is in the act of protest itself, not in the positive desire to change surrounding reality.

Authentic, indigenous authorities have been conspicuously absent in political and social life. Impatient for the effects of system transformation, people have expected not just an immediate improvement in living conditions, but positive changes in political culture and in the ethical posture of political elites as well. Public opinion surveys, however, have indicated consistently negative perceptions of political institutions and parties, as well as of individual politicians. This has produced a mistrust in the ultimate meaning and purpose of public activity: more than 76 per cent of people surveyed felt that those who wish to involve themselves in social causes are capable of achieving very little or nothing at all (Wyka and Gliński, 1992). Such pessimism and dislike of political elites and institutions has led to increased passivity, especially among young people, the group that is the most impatient and emotional.

Another factor of enormous significance has been the explicit marginalization of environmental and counterculture groups by 'Solidarity' political elites. This was already evident in opposition activity before 1989, when the leaders of underground Solidarity expressed virtually no interest in environmental issues. At most, they exploited these issues instrumentally. This attitude did not change after the opposition assumed power. Policies were formulated by elites behind closed doors. Ruling circles and the people associated with them treated the ecological movement as an unimportant, marginal phenomenon. For their part, representatives of the movement had virtually no contact with the political elites beyond the a few officials of the Ministry of Environmental Protection, officials who were, on the whole, leftovers from the old communist structures. There was no dialogue between the Solidarity government and environmentalists about controversial questions. On the contrary, old bureaucratic methods were employed: police units were dispatched against young people protesting the construction of a nuclear power plant in Żarnowiec and a dam at Czorsztyn (in 1989 through 1993).[4]

All this intensified the environmentalists' disillusionment with political institutions, radicalizing their views and their behavior. It is worth noting that of all the many Polish social movements and groups who engaged in public protest from 1989 to 1992, only ecological groups composed mainly of young people were met with decisive police and legal measures. This discrimination became glaringly obvious at Czorsztyn, where the local officials intensified civil and criminal proceedings against those who participated in the 1992 blockade. The authorities seem to treat a group fighting for general societal values, rather than merely for simple group interests, as *ipso facto* weak, an apt target of police repression. Such a tactical approach could set off a violent response on the part of the persecuted groups, radicalizing them to the point of 'soft' terrorism (which, generally speaking, is rather far from Polish traditions).

Greens in the Free Elections

Despite the fact that environmentalists trusted local self-government more than central institutions or political parties, ecological groups participated only sporadically in the local elections of 1990. In addition to the external reasons for their rejection of political involvement, structural factors internal to the movement were also important. The movement was dispersed, organizationally weak, and poorly coordinated. As a result, only a few members of ecology groups were elected to local government bodies.

Similarly, the first fully democratic parliamentary elections in post-war Poland, held in Autumn 1991, did not engage the ecological movement, although four political groups employing ecological rhetoric did participate in these elections. With the exception of the Polish Green Party, which formed at the end of 1988, these 'parties' were established and registered only a few months before the elections. They were ad-hoc, artificial entities lacking specific constituencies. Candidates for the Sejm and the Senate were frequently nominated through professional and/or personal contacts. The main goal of these 'ecological political parties' was to win parliamentary seats by exploiting trendy slogans and to satisfy their founders' personal political ambitions.

Some of these organizations attempted to use the organizational structure of such Communist-era organizations as the League for Nature Protection, the Polish Hunters' Union, the Polish Fisherman's Union, and so on. Others merged their electoral lists with existing parties that showed little interest in environmental issues, such as the right-wing Confederation for an Independent

Poland. The Polish Green Party formed an electoral coalition with one of these 'artificial' organizations, the Polish Ecological Party, after members of the Green Party created and registered the latter group in order to pre-empt the use of the acronym 'PPE' by other equally ambitious ecological pseudo-activists (Gliński, 1992).

The way in which these opportunistic parties were created and chose their candidates from outside the ecological movement disqualified them in the eyes of the movement's participants. For example, the oldest Polish ecological party, the Polish Green Party (PPZ), was perceived as tainted by its evident opportunism and by the presence in its ranks of former Communist party functionaries and suspected collaborators with the Communist Security Service. Constant internal organizational conflicts and factional divisions within the PPZ combined with its demagogic style and unconcealed eagerness for foreign financial aid to compromise the party in the eyes of most of the ecological movement's participants.

The ecological movement's lack of support for ecological pseudo-parties in the 1991 parliamentary elections was clearly a conscious act. The movement's participants simply viewed these parties as shady, opportunistic, and hard to get along with. At the same time, the movement's activists either rejected the possibility of their own political involvement out of hand, or they were discouraged by the daunting tasks of campaigning in parliamentary elections in light of the organizational, programmatic, and political immaturity of the movement.

The electoral defeat of all four 'ecological parties' (none of which won even a single parliamentary seat) was determined not only by the lack of support from the ecology movement but also by the style of campaigning they used: vague, promise-everything election platforms; simple-minded didacticism; a lifeless campaign style; an inability to define their own practical accomplishments, as revealed in embarrassing television interviews; and candidates' lack of strong, individual personalities.

This inept electoral campaign wasted a unique chance for electoral success by 'shortcut'. Even without a party base or support from a social movement, it should have been possible to appeal directly to the public imagination, which was already positively disposed to ecological issues. Ecological values have been salient since the period of radically open public life during the first phase of Solidarity's existence in 1980 and 1981, when environmental issues were exploited in the struggle against the Communists. Later, they surfaced explicitly in the public reactions to the Chernobyl catastrophe and the protests mounted by various ecology groups. Over the course of the decade Poles ceased to view ecological values as 'alien' and started to treat them as 'ours'. This, of course, is a necessary pre-condition for the political effectiveness of ecological rhetoric.

Prior to the elections, socially accepted ecological values had not been compromised or exhausted. They were fresh and novel, a particularly rare trait on the Polish political scene. Moreover, a peculiar trait of ecological values was their paradoxically neutral character in the public consciousness. Compared to other political options which may have possessed pejorative connotations or have been compromised in some way, ecology was not perceived as an ideology 'against something'. It was rather to a significant degree seen as an ideology 'for something' (despite the radicalism of some pro-environmental groups). The prevailing image of ecology was of something pleasant, harmless, and not entailed in political manipulations. It was therefore attractive, even inspiring, to broad sectors of society.

The Greens' political impact in Western democracies was noted favorably by the Polish electorate, although the perception of Western Greens as crypto-communists presented some problems.[5] However, as the process of transformation continues to unfold, we can anticipate greater acceptance for what might be called left-wing views. Though it should be noted that this process may also prove unfavorable for the Greens. If an authentic social democratic movement were to emerge soon in Poland, it would compete with them for the same electorate, as has already been the case in some Western democracies.

Finally, a peculiar characteristic of Polish society's environmental consciousness is its not-inconsiderably irrational, magical-fearful character, as was apparent in the aftermath of the Chernobyl disaster. Leaving aside moral valuations, a capable and intelligent socio-technician could exploit this attitude for electoral purposes.

It is worth noting that the observation about the attractiveness of environmental slogans and values finds confirmation in public opinion surveys of attitudes about political parties. Greens generally won the approval of between 10 and 40 per cent of those questioned. In surveys prior to local self-government elections in 1990, ecological symbolism won second place after 'Solidarity' slogans. Ecological parties also won the largest 'net' acceptance (the difference between support for a given party and the lack thereof) in 1991 pre-election surveys, with more than 40 per cent.

In summary, then, I would suggest that during the 1991 Polish parliamentary campaign specific psycho-social conditions appeared to favor electoral success for environmental parties. This went beyond a popular disposition toward ecological issues. In the electoral race itself, there was an absence of real political parties, group identities in society were blurred, old symbols were exhausted, and many social groups were disoriented. In other words, a vacuum in the political sphere favored the appearance of new, largely unfamiliar actors. But this opportunity for political success was completely squandered by pseudo-ecologists.

In 1993 no ecological parties ran for parliamentary election in Poland.[6] Formally registered Green parties are still largely artificial and have virtually no contact with the environmental movement.[7] The movement remains mistrustful of politics, despite many debates about necessary involvement in political arena. It is still too weak for organized electoral action and preoccupied with developing its own organizational structure. A real political involvement will probably be the next step in the movement's development in the future.

The Ecological Movement As an Element of Civil Society

Despite its organizational weaknesses and lack of widespread support, the environmental movement has developed rather quickly: new ecological groups and organizations are springing up, and their membership is growing. Organizational reforms are continuing, as is the professionalization of nongovernmental environmental institutions. The movement plays an important role in the process of systemic transformation toward the emergence of a civil society.

Civil society is characterized by the functioning of a richly-differentiated structure of group interests. These groups are organized from below, spontaneously, on the basis of independent choices made by citizens, and function independently of state structures. They are therefore capable of maintaining and defending their own autonomy. It is essential that a given group be able to increase its activity and mobilize when confronted with danger. The level of spontaneity and energy of interest groups change over time, usually growing stronger during the formative stages, in time of danger, during programmatic and tactical struggles (such as entrance into coalitions, or in the midst of changes in the social, economic, and political environment), and during the course of institutionalization. Individual group identities change and differences and cleavages among them become clarified. Civil society, then, can be defined as a market of interests, of which, however, only some interests have a purely economic character derived from the economic market. Many other interests and the social institutions associated with them are based on non-economic criteria such as value-orientations, ideology, or social, religious, cultural, and political criteria.

The 'interest market', like the narrower economic market, operates by means of an 'invisible hand': subjects pursuing their own particular interests help attain general social goals, thus contributing to the general good. This process determines the 'civility' of interest groups' action. 'Civility' here means the consent for the functioning of certain ground rules to which all are subject, such as legal regulations, parliamentarism, representative democracy, and social pacts. These rules are guaranteed primarily by the state, but are originally acknowledged and accepted by society (a fact with particular significance in conflicts between the state system and civil society).

According to Philippe Schmitter (in O'Donnell et al., 1986), 'interest group public status' is essential to the effective functioning of civil society because it entails social consent for public action, including the defense of group interests. 'Public status' endows interest groups with a 'citizenly' character, with a mandate to act relatively independently vis-à-vis the state authorities and even to refuse compliance under conditions of an autocratic government. Civil society is also characterized by specific modes of behavior and civic attitudes that emerge primarily in the course of the formation of group identities and social bonds.

Civil society can also be defined as a set of ethical norms for individual behavior. In addition to basic economic attitudes such as entrepreneurship and honesty and in addition to respect for the law, this 'civic posture' includes the following:

- the ability to function publicly (knowledge of how to form associations, a sense of group identity, participation in the administration of authority, control of one's affairs);
- tolerance and a willingness to acknowledge and negotiate with other interest groups;
- an understanding of what it means to be represented politically and acceptance of democratic rules (responsible action vis-à-vis the state); and
- the possession of so-called 'civic virtues', including a willingness to think and act for the common good.

The presence of a certain set of 'civic attitudes' is not, in itself, sufficient to make a society 'citizenly'. Usually, however, 'citizenliness' can be considered incremental. One can thus speak of a more or less civic attitude and about a more or less civic society.

It must be admitted that civil society in this sense is still far from fully realized in Poland, though it is one of the goals of the systemic transformation. Public activity by individuals and groups remains, generally speaking, limited. Group identity and horizontal bonds at the middle level are very weak, and the new democratic institutional shell has not yet been filled with social substance. However, one hopeful sign is the environmental movement, which represents a step in the direction of civil society. It is a relatively mature interest group possessing both a unique and well-defined identity (during the Communist era, a negative identity formed by opposition to the system predominated) and a significant degree of independence from state and political structures (during the period of 'underground civil society', environmental activists were also independent of the mainstream opposition). The movement arose from below, a result of individual decisions to participate. It is differentiated institutionally, able to organize itself, and able to defend its interests through social activism. The ecology movement has a specific public status (that is, social acceptance of its activities) and social bonds, internally and between groups. With some exceptions, it has demonstrated responsibility and loyalty vis-à-vis the state's fundamental rules. Finally, it has achieved a significant level of tolerance for other groups and organizations, and the ability to negotiate with them. The movement possesses basic 'civic virtues'; among them a positive value, which is not something to be ashamed of (though it is a common attitude since the fragmentation of the Solidarity-opposition).

In sum, the ecology movement is maturing in 'citizenliness', which means that it has begun to politicize. Basic forms of social action, such as the articulation of group interests and values and protests over grievances, are increasingly being superseded by more mature, typically 'civil' ways of acting: participation in the democratic process (cooperation with local administrative bodies, participation in legal means of conflict resolution, lobbying), as well as independent cultivation of the 'non-state' social sphere such as ecological and educational campaigns.

It needs to be noted that the 'civil' image of the environmental movement presented here does not apply to all nongovernmental organizations and informal groups. Certain phenomena within the movement threaten to disrupt its 'civility', while some otherwise 'civil' traits can lead to conflict. For example, the independence and subjectivity of the movement's participants clash with other civil traits, such as the formation of an identity for the entire movement, its coordination, and its institutionalization.

The ecological movement has shown a range of different attitudes toward the rules regulating the market of group interests. A radical minority rejects these rules altogether. This radicalization, characteristic of 'new social movements', is symbolized by the slogan: 'there can be no compromise in defense of Mother Earth'. Although radicals until now have refrained from aggression and violence, there continues to be a real danger of soft ecological terrorism, and the radicalization of attitudes within the ecological movement continues to be real, particularly in light of the marginalization and harassment of some groups.

The Polish ecological movement enjoyed a public status and social acceptance during the period of 'real existing socialism' and during the initial period of systemic transformation. Afterwards, there has been less and less social acceptance of extremist groups of all stripes, including radical ecologists. But this applies to their form of activism, not to the content of their program. If they incorporate the assumption of 'citizenliness' into their programs, that is, the principle of not infringing on other people's interests, it is possible that even the values of 'deep ecology' may gain acceptance. The cultural, value-oriented (often quite radical) character of the ecological movement reinforces 'civil virtues' regardless of the circumstances. But radical postures can lead to arrogance and unruliness, and interfere with broadly defined citizenly values. For instance, the acceptance of rules valid for all can be perceived as accommodationism and selling out. The ecological movement has a strongly ethical aspect, which can be regarded as a characteristic feature of civil society. Activists frequently view the goals and interests of their movement as universally valid; and therefore often seek to impose ecological values on society, whether or not they are appreciated as such. The temptation of ecological *diktat* and adoption of 'un-citizenly' methods is a present danger, although, until now, Polish Greens have succeeded in avoiding it.

Despite these dangers, the system of civic values and the types of social action represented by the environmental movement are genuinely valuable, even more so because they appear so rarely in contemporary Polish society. Looked at this way, the peculiar, dual educational function performed by the ecological movement towards its members and towards 'the outside' is particularly significant.

Within the movement, the existence of a broad and transparent market of group interests encourages and facilitates the spread of citizenly attitudes such as tolerance, negotiating skills, and responsibility vis-à-vis rules governing public conduct. This constitutes a self-reinforcing process, a mutual link between a developed interest group structure and civil attitudes, two crucial elements of civil society.

Regarding society at large, cultural models and social norms developed in the areas of civil action are gradually influencing behavior and postures in other areas of societal life. The activities of nongovernmental organizations and social movements encourage individual participation in public and political activity.

Social movements in present-day Poland fulfil one other important function, which follows from the necessity of making deep and rapid systemic changes, the inefficiency of democratic institutions, the difficult economic situation, post-Communist social consciousness, the disintegration of value systems and norms, and the immaturity of political elites. All these factors hinder the formation of civil postures and procedures. Under these conditions, social and cultural movements (including the ecological movement) represent a niche or incubator of citizenly activity acting to preserve civil values until the completion of the system's institutional reconstruction. This protects these values during a period in which they are especially vulnerable to the drastic economization of social life, with its associated social pathologies and cultural chaos.

Concluding Remarks About the Greens in the late 1990s

Generally speaking, the process of the slow evolution of the environmental movement observed nowadays in Poland appears similar to changes that occurred in the structure and activity of the social movements in the West after the 1960s. All these positive changes were made possible by favorable transformative reforms in the socio-political framework, particularly by the institutionalization of freedom of socio-political activity, the decentralization brought about by local government reform, and relatively stable conditions for social activity. Any future changes in these factors can hinder or even block the process of maturation of the ecological movement.

To a large extent, the Polish environmental movement reflects the features defining a social movement: it is characterized by a high degree of interaction among individuals and groups, and it is actively engaged in social and cultural change.

The professionalization, institutionalization, and 'politicization' (that is, cooperation with governmental and self-governmental bodies at the local and central levels) of environmental NGOs often keeps abreast of the spontaneously emotional activities and deep-ecological background of these organizations. The appropriate balance between the movement's level of organization and its ability to act spontaneously is maintained, enabling it to function effectively and to continue its development. The 1980s were dominated by a high degree of informality and a low level of organization, whereas the transformation process of the 1990s has been accompanied by the movement's increasing attention to organizing itself more formally. The maturation processes described here are based on the development of a variety of formal structures assumed by groups within the movement, resulting in an overall increase in its effectiveness. Yet the movement's ability to act spontaneously and flexibly is maintained by the variety and diversity of its organizational structures and the collective character of the movement's organizations. Increasing organizational development may endanger the environmental movement's identity and independence by, for instance, alienating its founders, expanding reliance on outside funding sources and/or government authorities for support, digressing to non-essential issues, etc.

One of the main features of the contemporary Polish environmental movement is still the differentiated character of its participants. Another very important new feature is a mutual exchange of experience and attitudes about ecological activity among various groups and organizations. Typical professionals become more and more sensitive to deep-ecological and countercultural ideas, or at least can tolerate them, which signifies obvious progress. 'Veterans' of deep ecology cooperate with local governments and are ready to make compromises with technocrats. For instance, the Workshop for All Beings, a group of a long deep-ecological tradition, has decided recently to make a deal with local authorities concerning the construction of a new, modern-type chair-lift at Pilsko mountain, which is to replace a few existing, obsolete ski-lifts, dangerous for nature.

As a result, the ecological movement is composed of various organizations using in their activity a specific mixture of different forms of acting, forms of activity that are above all flexible and pragmatic. The movement participants have learned this flexibility and pragmatism during their voluntary work in NGOs. This experience has been a decisive factor shaping the changes of the movement's character in Poland.

One of the dominant forms of NGOs is still a collectivist organization, a characteristic feature of other new social movements (Klandermans, 1989). The environmental groups in Poland are also a more inclusive than exclusive type of social movement organization (SMO), which means group identities become vague. The general identity of the movement, on the other hand, has even been reinforced by the 'exchange' of the group identities within the movement, and by the 'common enemy syndrome' caused by all participants of the movement sharing the general feeling of having such an enemy.

The Polish environmental movement is also largely characterized as alternative and unconventional. This otherness of the movement, compared to the mainstream values dominant in society, is seen in the issues it addresses as well as in the many types of activities it conducts. This alternative content and action is illustrated best by 'deep ecology' groups and youth-dominated animal rights groups.

In the process of developing, the movement's members have assumed an increasing level of self-awareness. In the 1980s the movement's identity was shaped by an awareness of belonging to a community that was above all else formed as a negative reaction to a common political opponent. During the period of systemic transformation the development of this awareness is undergoing a slow change, and is now defining itself as a reaction to the dominant society. However, growing numbers of shared experiences and 'positive' activities, especially through the processes of 'self-education' and 'maturation', are also contributing to the formation of this self-awareness. Activities integrating and unifying the movement play an important role in the creation of this self-awareness. These activities include establishing information and communication networks, institutionalizing membership meetings, and conducting joint activities and environmental campaigns. All of these contribute to reflection on the part of the movement's members, to the development of collective meanings, and ultimately to the creation of the movement's identity.

In sum, the Polish environmental movement possesses the organizational character and cultural background of a typical new social movement of the West, despite the differences in their origins. But at the same time, the Polish movement is relatively pragmatic and flexible enough to increase its chances of success. Its field of activity make it a typical new social movement: it is addressing 'new social issues', and seeks to act on global and universal values (albeit to a lesser degree than in the West because of the extent of environmental damage in Poland). It is consciously reluctant to engage in politics and operates in the sphere of noninstitutionalized policy (Offe, 1985). It adds to the classical mechanisms of democracy the spirit and efficiency of public participation and grassroots democracy.

The Polish independent environmental movement was initiated in 1980 with the changes brought about through the Solidarity trade union movement. As a protest movement, it was actively engaged in the struggle for social and political change during the 1980s. After 1989, the movement in general is not opposing the political system, but is involved above all in working for change of a cultural nature. The movement is first of all a cultural phenomenon, functioning in the sphere of noninstitutionalized politics. In Poland it personifies a 'new' type of social conflict, between two concepts of culture, between two styles of life, two ways of organizing society, two symbolic spheres. The conflict between dominant and alternative cultures is played out in everyday life and in the sphere of civil society.

The dilemmas emerging in the Polish environmental movement are typical of new social movements. There is a certain tension between the counterculture character of its activities and its ultimate goals, which can be achieved only through mainstream politics. One part of the movement may be shouting uncompromising slogans and conducting radical actions, while in contrast another segment adopts a more tempered style marked by rational and self-imposed limitations. The universal values of environmentalism are not always reflected in the limited content and scale of the activities conducted by many of the movement's groups. One group is inclusive, while another may be closed to all but its self-appointed inner circle. Some groups are very active and creative, others are passive, turn defensive, or assume a utopian-moralistic approach. Proposals to reject modern materialism are found next to the promotion of current alternative technologies. Finally, the vision of the future offered by the movement seems to have little chance of success given Polish conditions, less chance, at any rate, than in the West.

The environmental movement plays a number of very specific roles in the process of systemic transformation in Poland. Above all, it is one of a very few grassroots agents for change with an evolutionary character. It assumes the role of prophet, in contrast to leading politicians expounding a desirable vision of the future for the country. Systemic transformation enabled the environmental movement to change its character in Poland, moving from a political opposition movement to a socio-cultural movement, active in the sphere of noninstitutionalized politics. It is most likely that the movement will advance to the next stage of its development, active involvement in institutionalized politics. The length of time a system undergoes liberalization plays a key role in the processes of maturation of a social movement and the development of its civic nature. This period of liberalization prepares the cultural foundation for the institutional transformations which follow. The quality of the social changes inspired by the Polish environmental movement is not only 'pro-environmental' but also to a large degree civic. The environmental movement is an enclave of civil society and promotes its development on a wider scale. Additionally, the movement 'preserves' the values of civil society during a time when increased social tension results from systemic transformation. As a citizen-based (civic) and self-limiting movement, it also contributes to the stabilization of the system and mitigates the forms of social conflict during the period of change. The Polish political elites in implementing their version of systemic change from above do not support the development of the environmental movement. Sharp social conflicts that characterize periods of systemic transformation do not promote the broader development of 'new social movements' or citizen initiatives, despite the fact that these movements often offer interesting proposals for reform and a certain potential for action. Alternative programs promoting socio-cultural changes are rejected by society at-large, primarily because the ordinary citizen cannot understand or accept the long time-line needed before these programs' effects are felt. In this situation, the environmental movement cannot become a mass movement and must remain at the margins of society, all the while yet fulfilling vital social roles such as the development of cultural and social norms, thus contributing to the broader goals of transformation.

The dynamic of change in the Polish environmental movement during the period of systemic transformation can be expressed by a number of distinctively dominating processes. This movement is on the road: from opposition to citizen participation; from an elitist, isolated counterculture to civil society; from the emotive to the pragmatic; and, finally, from informal spontaneity to mature institutionalization and professionalization.

Notes

1. Without taking into consideration some local branches of P.E. (Polish Ecological Club), LOP (League of Nature Protection) and 'Animals' Foundation. According to data provided by the Polish Information Bank of Non-Governmental Organizations KLON/JAWOR there are also about 700 NGOs concerned with ecological and other problems (declaring ecological goals among other aims of the organization), and environmental organizations amount to 14-17 per cent of the Third Sector in Poland. The whole Polish Third Sector, according to different estimates, accounts for 17000- 47000 organizations. (JAWOR, 1995).

2. The intelligentsia is considered here as a specific social group characterized by its unique ethos of mission and civic virtues as well as rationality and self-restraint.

3. The Democratic Union later merged with another new, pro-reform party, the Liberal Democratic Congress, to become the Freedom Union, which is currently a minority party in the Polish Parliament.

4. The government eventually abandoned plans for the nuclear power plant at Żarnowiec, but the Czorsztyn dam was completed despite the movement's protests.

5. It is impossible to place Polish environmental groups on a binary scale of political views. Divisions in left and right — as on the entire Polish political scene — are impossible to correlate in the case of 'ecologists'.

6. A few participants of the movement won seats in the Parliament as members of the 'Ecological Forum' of the Freedom Union.

7. There were 17 'green' parties of this kind at the end of 1995. All of them were typical 'sofa' parties — all members of the party could fit on one sofa — without any support from the movement.

References

Devall, Bill and George Session (1985), *Deep Ecology: Living as if Nature Mattered,* Peregrine Smith, Salt Lake City.

Eder, K. (1993), *The New Politics of Class. Social Movements and Cultural Dynamics in Advanced Societies*, Sage Publications, London.

Gliński, Piotr (1992), 'Zieloni na politycznej scenie' (Greens at the political arena), *Krytyka*, No. 38.

Gliński, Piotr (1996), *Polscy Zieloni. Ruch społeczny w okresie przemian (Polish Greens: A Social Movement in the Time of Transformation)*, Wydawnictwo IFiS, Warsaw.

JAWOR (1995), *Informator o organizacjach pozarządozych*, JAWOR, Warsaw.

Klandermans, B. (ed.) (1989), *International Social Movement Research: Organizing for Change*, Vol. 2, JAI Press, New York.

Kolarska-Bobińska, Lena (1990), 'Civil Society and Social Anomie in Poland', *Acta Sociologica*, No. 4, pp. 277-288.

O'Donnell, G., P.C. Schmitter, and Whitehead, L. (eds), (1986), *Transition from Authoritarian Rule. Southern Europe*, Johns Hopkins University Press, Baltimore.

Offe, C. (1985), 'New Social Movements: Challenging the Boundaries of Institutional Polities', *Social Research*, No. 4..

Wygnański, J. J. 1995a. Raport o organiacjach — Public Policy Oriented, Warsaw.

Wygnański, J. J. (1995b), 'Czym jest Trzeci Sektor w Polsce?' [What is the Third Sector in Poland?], paper for the seminar 'Building Civil Society: Role of Nongovernmental Organizations in Poland', Republic of Poland, Senate, 12 June, Warsaw.

Wyka A., Gliński P. (1992), *Swiadomość ekologiczna społeczeństwa polskiego (Ecological consciousness of the Polish society)* komunikat z badań (report from the survey) OBOP, Warsaw.

Zald M. N., and McCarthy J. D. (1990), *Social Movement, in an Organizational Society: Collected Essays*, Transaction Publications, New Brunswick.

Index

References from Notes indicated by 'n' after page reference

Balcerowicz, Leszek, 6
Bochniarz, Zbigniew and Richard Bolan, 5, 6, 10, 11, 12
budget constraints, 7, 8, 11, 14
  hard under capitalism, 72
  soft, contribute to pollution under socialism, 30, 71

Chernobyl disaster, 111n, 142, 143
civil society, 6, 8-9, 11-12, 14, 148-9
  civility and 'citizenliness', 144-6
  interests within, 10, 144-5
  environmental movement as component of, 144-8
  'command-and-control' regulations, 7, 31, 96-7
Cummings, Susan, 5
Czorsztyn dam, 131, 135, 141

debt-for-nature swaps, 24-5, 47, 49-50
deep ecology, 130, 131, 149
democracy, 6, 10, 14

EcoFund, 19, 25, 47, 48-49
environmental fees and fines, 10, 20-1, 22, 26, 30-67, 122-3
  among highest in the world, 31, 34, 37-8, 50, 93
  below Pigouvian levels, 34, 55
  low collection rate for fines, 21, 26, 34, 36, 53, 54
  product fees, 43, 55, 58
  trade-off between incentive and revenue raising role, 52-3
  weaknesses and problems, 54-9
environmental funds, 12, 19-20, 22-4, 45, 95
  decentralization of management, 22, 28n, 55
  Farmland Protection Fund, 19, 23-4
  Forestry Fund, 19, 23-4
  Local Funds for Environmental Protection and Water Management, 19, 20, 22, 23, 27, 32-3
  National Fund for Environmental Protection and Water Management, 19, 20, 22-3, 24, 27, 32-3, 38, 55, 74, 83-4
  Voivod [regional] Funds for Environmental Protection and Water Management, 20, 22, 23, 27, 32-3
environmental impact assessments, 93-4, 95, 97
environmental investments, 12-13, 23, 32, 34, 45-9, 54, 125-6
environmental legislation, 81-115, 122
  communist era, 5, 7, 9, 82-90
  during transition, 10, 91-6, 100-111
  still to be adopted, 96-100
environmental loans, 5, 12-3, 22, 23, 24, 27, 45, 93
environmental movement, 3, 8-9, 16n, 86, 129-53
  institutionalization, 148, 150-2
  maturation, 134-8
  politicization, 135-6, 138-41, 146
  professionalization, 133-4, 136-7
  self-education, 134, 149-50

environmental policy community, 4-5, 84-5, 89
Environmental Protection and Development Act of 1980, 6, 7-8, 10, 12, 16n, 81-111 *passim*, 112n
Environmental Protection Bank, 14, 19, 20, 24
environmental protection budget, 20
  protected from political interference, 12, 19, 26-7, 32, 101
environmental taxes, 20, 38-9, 43-5, 77
European Union, 13, 15, 125
  PHARE program, 78, 91, 136
  setting regulatory standards for aspiring members, 61-2, 81, 91, 93, 96, 111, 125

Fiedor, Bogusław, 4, 7
'Freedom and Peace' movement, 131, 135, 138

Garton Ash, Timothy, 6
Gliński, Piotr, 5, 8, 11
'green business', 11, 26, 98-99

Hetman National Investment Fund, 124-5

Information Bank of Nongovernmental Organizations (KLON/JAWOR), 134, 152n

Jancar-Webster, Barbara, 3
Jendrośka, Jerzy, 4-5, 6, 7, 8, 9, 11, 12

Manser, Roger, 2
market-based environmental policies, 5, 7-8, 10, 12-3, 19, 21, 30-69, 72-79, 99-100
market economy in Poland, 6, 10, 14, 26
Martial Law (1981), 8, 86
Mass Privatization Program, 5, 116-28 *passim*
  environmental management, 77-8, 121-7
  methods of privatization, 116-8
Mazowiecki, Tadeusz, 138
Ministry of Environmental Protection, Natural Resources and Forestry, 3, 10, 12, 15, 21, 24, 43, 45, 75, 83, 135
Ministry of Privatization, 5

National Environmental Policy of 1990-1, 3, 30, 91
National Investment Funds, 5, 117-27 *passim*
Nature Conservation Act of 1991, 92
nongovernmental organizations (NGOs), 9, 11, 12, 13, 22, 84-89, 108, 130-154 *passim*
Nowicki, Maciej, 25

Paris Club, 24-25, 49
Polish Ecology Clubs, 8, 87, 131, 152
Polish Environmental Law Association, 9, 16n, 85
Polish Green Party, 138, 141-142
'polluter pays principle', 20, 28n, 31, 74-8, 83
  and environmental liability, 36, 73-6
  and 'user pays principle', 36
pollution
  in communist Poland, 1-2, 16n
  and firms' operating permits, 21, 31, 36, 122-3
  and public health, 1-2
  reductions during transition, 3-4, 15, 16n, 25
  transferable emission quotas, 54, 56-7, 91
pricing, 8, 11, 20, 37
privatization, 5, 13, 14, 68-80, 103, 108, 116-28
public health, 1-2
public participation, 8-9, 84, 88-9, 105, 112n, 129-53 *passim*

Rakowski, Mieczysław, 6, 9
regulatory conflict of interest under socialism, 10-11, 72, 77
'Round Table' accords, 9, 91
rule of law, 6, 10

Schmitter, Philippe, 145
'shock therapy', 6, 19, 60
Śleszyński, Jerzy, 4, 6, 12
socialism,
  causes of heavy pollution, 8, 30, 69-71, 88-91
  cultural heritage as barrier to political activity, 139-140
Solidarity, 6, 8, 9, 12, 129, 142, 143
  contribution to rise of environmental movement, 129
  marginalization of environmental groups by, 140
State Environmental Protection Inspectorate, 11, 15, 21-22, 36, 51, 95-6, 114n
  regional inspectorates, 21, 36
sustainable development, 26, 70, 76-79, 96-7, 132

United States Agency for International Development, 5, 136

Wałesa, Lech, 6
Western assistance, 24-5, 49-50, 91, 136
World Bank, 15

Żarnowiec nuclear power plant, 131, 135, 141
Żylicz, Tomasz, 28n

For Product Safety Concerns and Information please contact our EU representative GPSR@taylorandfrancis.com Taylor & Francis Verlag GmbH, Kaufingerstraße 24, 80331 München, Germany

Batch number: 10726007

Printed by Printforce, the Netherlands